Mast(
AP® Statistics

Gerry McAfee

Course Technology PTR

A part of Cengage Learning

COURSE TECHNOLOGY
CENGAGE Learning™

Australia • Brazil • Japan • Korea • Mexico • Singapore • Spain • United Kingdom • United States

COURSE TECHNOLOGY
CENGAGE Learning™

Master Math: AP Statistics

Gerry McAfee

Publisher and General Manager,
Course Technology PTR:
Stacy L. Hiquet

Associate Director of Marketing:
Sarah Panella

Manager of Editorial Services:
Heather Talbot

Marketing Manager:
Jordan Castellani

Senior Acquisitions Editor:
Emi Smith

Project Editor: Dan Foster,
Scribe Tribe

Technical Reviewer: Chris True

Interior Layout Tech:
Judy Littlefield

Cover Designer: Jeff Cooper

Indexer: Larry Sweazy

Proofreader: Brad Crawford

© 2011 Course Technology, a part of Cengage Learning.

ALL RIGHTS RESERVED. No part of this work covered by the copyright herein may be reproduced, transmitted, stored, or used in any form or by any means graphic, electronic, or mechanical, including but not limited to photocopying, recording, scanning, digitizing, taping, Web distribution, information networks, or information storage and retrieval systems, except as permitted under Section 107 or 108 of the 1976 United States Copyright Act, without the prior written permission of the publisher.

For product information and technology assistance, contact us at
Cengage Learning Customer & Sales Support, 1-800-354-9706
For permission to use material from this text or product,
submit all requests online at cengage.com/permissions
Further permissions questions can be emailed to
permissionrequest@cengage.com

AP, Advanced Placement Program, College Board, and SAT are registered trademarks of the College Entrance Examination Board.

All other trademarks are the property of their respective owners.

All images © Cengage Learning unless otherwise noted.

Library of Congress Control Number: 2010922088

ISBN-10: 1-4354-5627-0

ISBN-13: 978-1-4354-5627-3

Course Technology, a part of Cengage Learning
20 Channel Center Street
Boston, MA 02210
USA

Cengage Learning is a leading provider of customized learning solutions with office locations around the globe, including Singapore, the United Kingdom, Australia, Mexico, Brazil, and Japan. Locate your local office at:
international.cengage.com/region

Cengage Learning products are represented in Canada by Nelson Education, Ltd. For your lifelong learning solutions, visit courseptr.com
Visit our corporate website at cengage.com

Printed in the United States of America
1 2 3 4 5 6 7 12 11 10

Table of Contents

Acknowledgments

I would like to thank Chris True, AP Statistics teacher and consultant, for reading and editing this book for content. I appreciate his helpful comments and suggestions. I would also like to thank Dan Foster for all of his efforts in editing this book. Dan was extremely easy to work with and provided me with excellent feedback. I am grateful to Judy Littlefield for all of her hard work on the many illustrations and equations.

I extend my thanks to Emi Smith, Senior Acquisitions Editor, and Stacy Hiquet, Publisher and General Manager, for the opportunity to write this book. Their patience, support, and trust are truly appreciated. I would also like to thank Brad Crawford for proofreading. Additional thanks to Sarah Panella, Heather Talbot, Jordan Castellani, Jeff Cooper, and Larry Sweazy.

I am grateful to have had some great teachers while growing up. I want to specifically thank Sandy Halstead for teaching me algebra, algebra 2, trigonometry, and physics. I am grateful to Karen Ackerman for teaching me honors English.

To my students, past and present, thank you for motivating me to be the best teacher I can be. I am inspired by the efforts you make in preparing for the AP Exam. I hope you continue those efforts in all aspects of your lives.

Finally, I would like to thank my friends and family. To my friends—especially John, Chris, Brad, and Tim—thank you for all your support and encouragement. To my friend and department head, Dan Schermer, thanks for always guiding me in the right direction. To Dean and my "painting buddies," thanks for providing me many hours of fun and stress relief as we worked together. Special thanks to Nick and Chad for helping me with graphs and images. To Bob and Carol, thanks for all of your support. To all of my extended family, thank you for all of the help you have given to me, Lori, and our children. To my parents, Don and Joan McAfee, thank you for being great parents and helping me develop a strong work ethic and for instilling me with strong values. To my sister, Lynn McAfee, and her family, thank you for all of your encouragement and support. To my children, Cassidy and Nolan, thank you for being such great kids and working so hard in school. Your talents and hard work will pay off. I appreciate you constantly asking me, "Are you done with your book yet, dad?" This question motivated me to keep pressing on. And finally, to my wife, Lori, I could not have done it without you! Thanks for doing more than your fair share and always believing in me!

About the Author

Gerry McAfee began his teaching career as an undergraduate at Purdue University. He is currently teaching AP Statistics in Brownsburg, Indiana, and has been teaching mathematics for 17 years. In addition to teaching AP Statistics, Gerry also teaches dual-credit mathematics, including finite math, through Indiana State University, and applied calculus, through Ball State University. Gerry has also taught a wide range of additional math courses including basic math, pre-algebra, algebra, and algebra 2. Gerry has won teaching awards including the Brownsburg High School National Honor Society Teacher of the Year, Who's Who Among American Teachers, and the Fellowship of Christian Athletes–Positive Role Model. He was also chosen for teacher recognition by an *Indianapolis Star* Academic All Star student.

Gerry graduated from Purdue University with a Bachelor of Science degree in Mathematics in 1993. He also holds a Master of Arts degree in Education from Indiana Wesleyan University.

In his spare time, Gerry enjoys spending time with his wife, Lori, and two children, Cassidy and Nolan. Family activities include many hours of school functions and sporting events, along with family vacations. In the summer, Gerry enjoys spending time with his teaching colleagues outdoors painting houses. Other activities that Gerry enjoys include fishing, running, and most recently triathlons. Gerry has run seven marathons and has run the Boston Marathon two times (not in the same day).

Introduction

AP Statistics is part of the *Master Math* series. The series also includes *Basic Math, Pre-Calculus, Geometry, Trigonometry,* and *Calculus.* This series includes a variety of mathematical topics and should help you advance your knowledge of mathematics as it pertains to these subjects.

AP Statistics is written specifically with you, the AP Statistics student, in mind. All topics of the AP Statistics curriculum are discussed within the 10 chapters of this book. These topics are explained in a manner suitable for a wide variety of ability levels. The topics are arranged so that you can develop an understanding of the various concepts of AP Statistics, including *Exploring Data, Sampling and Experimentation, Anticipating Patterns,* and *Statistical Inference.* All example problems in each chapter include solutions with the AP Statistics Exam in mind. These solutions will help you understand not only the concepts at hand but also how to communicate that understanding effectively to the reader (grader) of the exam.

AP Statistics includes some useful appendixes. All tables given on the AP Statistics Exam are included in Appendix A. Appendix B includes all formulas needed for the AP Statistics Exam. These formulas are separated into two categories: those formulas that are given on the exam and those that are not. You will also need to have a good understanding of the "assumptions and conditions" for inference. These "assumptions and conditions" are fully discussed within each chapter that deals with inference and are summarized in Appendix C for quick reference.

A glossary is included at the end of the book as well so that you can reference or study any vocabulary terms.

The following section, "Preparing for the AP Statistics Exam," is written to help you develop a sense of how the AP Statistics Exam is organized as well as how best to prepare yourself for the upcoming examination.

Although *AP Statistics* may not totally replace your textbook, it should provide you with some great insights on how to tackle the types of questions that you will likely encounter on the AP Exam. It is a comprehensive book that should prove to be an invaluable resource as you journey toward your goal of reaching your maximum potential on the AP Statistics Exam. Good luck!

Preparing for the AP Statistics Exam

- Preparing thoroughly for the AP Statistics Exam is essential if you wish to perform well on the exam and earn a passing grade. Proper preparation begins the first day of class. Like anything worthwhile in life, reaching your potential on the AP Statistics Exam takes hard work and dedication.

Plan for Success

- Get motivated! Begin your preparation for the AP Statistics Exam early by doing *all* of your homework on a daily basis. Doing "some" or even "most" of the work is selling yourself short of what you are capable of achieving. Manage your time and get *all* of your work done. You may find AP Statistics to be more a little more difficult than some other math courses you have taken, and there might be an adjustment period before you are achieving at a high level. Be patient and keep working!

• Do all of the reading assignments you are assigned. If your instructor does not assign you to read this book or your textbook, take it upon yourself to do so. Not only will you learn the material, you will also strengthen your reading comprehension and your ability to write well, which is important on the free-response portion of the exam. It is imperative that you can read and interpret questions effectively in order for you to understand the information being given and what you are being asked to do for each problem. Discipline yourself to keep up with your daily work, and do the appropriate reading!

• Review on a weekly basis. Even a few minutes a week spent reviewing the topics you have learned previously will help you retain the material that you will be tested on during the exam. If you do nothing else, study the glossary of this book, as it will keep your vocabulary of AP Statistics up to par. You might find it useful to review old tests and quizzes that you have taken in class. If you're like most students, you are very busy, and you'll need to really budget your time in order to review weekly *and* keep up with your daily work in this and other courses. It sounds easy enough, but again, it takes discipline!

• The more you do throughout the course, the easier your review will be toward the end of it. Realize, however, that you'll probably need to do a lot of review in the final weeks leading up to the exam. I recommend that you do "focused," or "intense," review in the last three to four weeks leading up to the exam. Don't wait until two or three days before the exam. "Cramming" for an exam like the AP Statistics Exam is not a good idea. You will be tested on most if not all topics in some way, shape, or form. Again, keep up with whatever your teacher or instructor throws your way.

- You should know and understand everything in this book to the best of your ability. Read it and study it! Try to go back after you have read the material and do the example problems on your own. As mentioned earlier, know the glossary. It will help you know and understand the terminology of AP Statistics. Get help on any topics that you do not understand from your instructor.

- Do as many "released" exam problems as you can from the College Board's website. Your instructor may have you do these for review, but if not, get on the website and do as many problems as you can. There are free-response exam questions there from as far back as 1997, and you'll have plenty to choose from. This will give you a good feel for the type of problems you should expect to see on both the multiple-choice and free-response portions of the exam. You will also find it useful to read through the grading rubrics that are given along with the problems. Doing these "released" free-response questions will help you understand how partial credit works on the exam. I do, however, think it's more important to understand the concepts of the problems that are given and what the grading rubric answers are than how many points you would have gotten if you had answered incorrectly. But it's still worth a little time to think about how you would have scored based on your answer and the grading rubric.

- Making your review for the AP Statistics Exam something you do early and often will prevent you from having to "cram" for the test in the last couple of days before the exam. By preparing in advance, you will be able to get plenty of sleep in the days leading up to the exam, which should leave you well rested and ready to achieve your maximum potential!

How AP Grades Are Determined

• As you are reviewing for the AP Statistics Exam, it's important that you understand the format of the exam and a little about the grading. Knowing the format of the exam, reading this book, and doing as many old AP Statistics Exam questions will have you prepped for success!

• The AP Statistics examination is divided into two sections. You will have 90 minutes to do each section. The first section is the multiple-choice section of the exam, which consists of 40 questions. The second section of the exam is the free-response portion of the exam and consists of 6 questions. The scores on both parts of the exam are combined to obtain a composite score.

• The multiple-choice portion of the exam is worth a total of 40 points but is then weighted to 50 points. The score on the multiple-choice section of the test is calculated by using the following formula:

[*Number correct out of* 40 − (0.25 × *Number wrong*)] × 1.2500 = Multiple-Choice Score

• The adjustment to the number of correct answers you receive makes it unlikely that you will benefit from random guessing. If you can eliminate one of the choices, then it is probably to your benefit to guess. If you cannot eliminate at least one choice, do not guess; leave it blank.

• The free-response portion of the exam is graded holistically. For that reason, it is to your advantage to try every question, if possible. Even if you don't fully understand how to answer the question in its entirety, you should still try to answer it as best you can. Scores on individual free-response questions are as follows:

4   Complete Response

3   Substantial Response

2   Developing Response

1   Minimal Response

0   No Credit

     No Response

• The AP readers (graders) grade free-response questions based on the specified grading rubric. If the question has multiple parts, each part is usually graded as "essentially correct," "partially correct," or "incorrect." Then, depending on the rubric, you will earn a 4, 3, 2, 1, or 0 for that particular question. Each score in the free-response question is weighted. Problems 1–5 on the free-response contribute 7.5 percent each to the maximum possible composite score, and question 6 contributes 12.5 percent. It is usually recommended in the directions of the free-response questions to spend more time on question 6 because it is worth more. Question 6 is considered an "investigative task" question and will probably require more in-depth thinking than the first five questions. Typically, you will be instructed to spend about 25 minutes on question 6. That will leave you with about 65 minutes to do the first 5 questions, which is about 13 minutes each.

• Once both parts of your test have been graded, a composite score is formed by weighting the multiple-choice and free-response sections equally. You will not be given your composite score. Instead, you will receive an AP Exam score based on the following 5-point scale:

5   Extremely Well Qualified

4   Well Qualified

3   Qualified

2   Possibly Qualified

1   No Recommendation

Don't worry too much about how the score is calculated. Realize that there are 40 multiple-choice questions that you must get done in 90 minutes and use your time accordingly. Don't spend a lot of time on a question that you find really difficult. Move on to the other questions and then come back to the difficult question(s) if time permits. Remember, if you cannot eliminate any of the choices, leave the answer blank. Also realize that you have 90 minutes to complete the free-response section of the exam. I recommend reading all six free-response questions quickly and starting with the one you think you have the best shot at answering completely and correctly. Be sure to read each question very carefully before you actually begin the problem. You don't want to invest a lot of time working on a problem and later realize that your answer doesn't really answer the question at hand.

• Make no mistake about it: The AP Statistics Exam is tough. You need to be ready. By reading and studying this book, doing your daily work on a regular basis, and doing old AP Statistics Exam questions, you will be properly prepared. Remember, the exam is designed to be tough, so don't get discouraged if you don't know how to answer every single question. Do your best! If you work hard at it and take it seriously, you'll leave the exam feeling good about yourself and your success. Good luck!

Chapter

1

Exploring and Graphing Univariate Data

1.1 Describing Distributions

• The organization of data into graphical displays is essential to understanding statistics. This chapter discusses how to describe distributions and various types of graphs used for organizing univariate data. The types of graphs include modified boxplots, histograms, stem-and-leaf plots, bar graphs, dotplots, and pie charts. Students in AP Statistics should have a clear understanding of what a variable is and the types of variables that are encountered.

• A variable is a characteristic of an individual and can take on different values for different individuals. Two types of variables are discussed in this chapter: categorical variables and quantitative variables.

Categorical variable: Places an individual into a category or group

Quantitative variable: Takes on a numerical value

Variables may take on different values. The pattern of variation of a variable is its **distribution**. The distribution of a variable tells us what values the variable takes and how often it takes each value.

Shape, Center, and Spread

• When describing distributions, it's important to describe what you see in the graph. It's important to address the **shape**, **center**, and **spread** of the distribution *in the context of the problem.*

• When describing **shape**, focus on the main features of the distribution. Is the graph approximately symmetrical, skewed left, or skewed right?

Symmetric: Right and left sides of the distribution are approximately mirror images of each other (Figure 1.1).

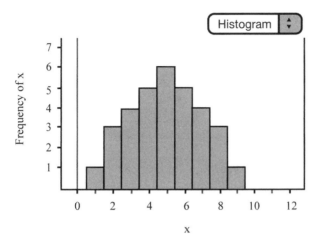

Figure 1.1 Symmetrical distribution.

• **Skewed left:** The left side of the distribution extends further than the right side, meaning that there are fewer values to the left (Figure 1.2).

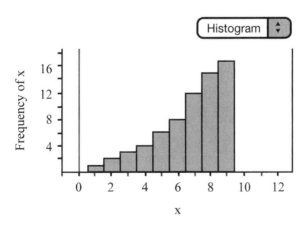

Figure 1.2 Skewed-left distribution.

• **Skewed right:** The right side of the distribution extends further than the left side, meaning that there are fewer values to the right (Figure 1.3).

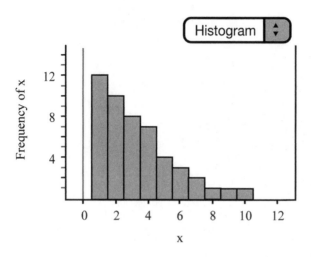

Figure 1.3   Skewed-right distribution.

• When describing the **center** of the distribution, we usually consider the mean and/or the median of the distribution.

Mean: Arithmetic average of the distribution:

$$\bar{x} = \frac{x_1 + x_2 + \ldots + x_n}{n} \quad \text{or} \quad \bar{x} = \frac{\sum x_i}{n}$$

Median: Midpoint of the distribution; half of the observations are smaller than the median, and half are larger.

To find the median:

1. Arrange the data in ascending order (smallest to largest).
2. If there is an odd number of observations, the median is the center data value. If there is an even number of observations, the median is the average of the two middle observations.

* **Example 1:** Consider Data Set A: 1, 2, 3, 4, 5

Intuition tells us that the mean is 3. Applying the formula, we get:

$$\frac{1+2+3+4+5}{5} = 3$$

Intuition also tells us that the median is 3 because there are two values to the right of 3 and two values to the left of 3. Notice that the mean and median are equal. This is always the case when dealing with distributions that are exactly symmetrical. The mean and median are approximately equal when the distribution is approximately symmetrical.

* The mean of a skewed distribution is always "pulled" in the direction of the skew. Consider NFL football players' salaries: Let's assume the league minimum is $310,000 and the median salary for a particular team is $650,000. Most players probably make between the league minimum and around $1 million. However, there might be a few players on the team who make well over $1 million. The distribution would then be skewed right (meaning that most players make less than a million and relatively few players make more $1 million). Those salaries that are well over $1 million would "pull" the mean salary up, thus making the mean greater than the median.

- When dealing with symmetrical distributions, we typically use the mean as the measure of center. When dealing with skewed distributions, the median is sometimes used as the measure of center instead of the mean, because the mean is not a *resistant measure*—i.e., the mean cannot resist the influence of extreme data values.

- When describing the **spread** of the distribution, we use the IQR (interquartile range) and/or the variance/standard deviation.

IQR: Difference of the third quartile minus the first quartile. Quartiles are discussed in Example 2.

Five-number summary: The five-number summary is sometimes used when dealing with skewed distributions. The five-number summary consists of the lowest number, first quartile (Q_1), median (M), third quartile (Q_3), and the largest number.

- **Example 2:** Consider Data Set A: 1, 2, 3, 4, 5

1. Locate the median, 3.
2. Locate the median of the first half of numbers (do not include 3 in the first half of numbers or the second half of numbers). This is Q_1 (25th percentile), which is 1.5.
3. Locate the median of the second half of numbers. This is Q_3 (75th percentile), which is 4.5.
4. The five-number summary would then be: 1, 1.5, 3, 4.5, 5.

- **Variance/Standard Deviation:** Measures the spread of the distribution about the mean. The standard deviation is used to measure spread when the mean is chosen as the measure of center. The standard deviation has the same unit of measurement as the data in the distribution. The variance is the square of the standard deviation and is labeled in units squared.

The formula for variance is:

$$s^2 = \frac{(x_1 - \overline{x})^2 + (x_2 - \overline{x})^2 + \ldots + (x_n - \overline{x})^2}{n-1}$$

or

$$s^2 = \frac{\sum (x_i - \overline{x})^2}{n-1}$$

The standard deviation is the square root of the variance:

$$s = \sqrt{\frac{(x_1 - \overline{x})^2 + (x_2 - \overline{x})^2 + \ldots + (x_n - \overline{x})^2}{n-1}}$$

or

$$s = \sqrt{\frac{\sum (x_i - \overline{x})^2}{n-1}}$$

- **Example 3:** Consider Data Set A: 1, 2, 3, 4, 5

We can find the variance and the standard deviation as follows:

Variance

$$s^2 = \frac{(1-3)^2 + (2-3)^2 + (3-3)^2 + (4-3)^2 + (5-3)^2}{5-1}$$

$$s^2 = 2.5$$

Standard Deviation

$$s = \sqrt{2.5} \approx 1.5811$$

It's probably more important to understand the concept of what standard deviation means than to be able to calculate it by hand. Our trusty calculators or computer software can handle the calculation for us. Understanding what the number means is what's most important. It's worth noting that most calculators will give two values for standard deviation. One is used when dealing with a *population*, and the other is used when dealing with a *sample*. The TI 83/84 calculator shows the population standard deviation as x and the sample standard deviation as S_x. A **population** is all individuals of interest, and a **sample** is just part of a population. We'll discuss the concept of population and different types of samples in later chapters.

• It's also important to address any *outliers* that might be present in the distribution. **Outliers** are values that fall outside the overall pattern of the distribution. It is important to be able to identify *potential* outliers in a distribution, but we also want to determine whether or not a value is *mathematically* an outlier.

• **Example 4:** Consider Data Set B, which consists of test scores from a college statistics course:

98, 36, 67, 85, 79, 100, 88, 85, 60, 69, 93, 58, 65, 89, 88, 71, 79, 85, 73, 87, 81, 77, 76, 75, 76, 73

1. Arrange the data in ascending order.

36, 58, 60, 65, 67, 69, 71, 73, 73, 75, 76, 76, 77, 79, 79, 81, 85, 85, 85, 87, 88, 88, 89, 93, 98, 100

2. Find the median (average of the two middle numbers): 78.

3. Find the median of the first half of numbers. This is the first quartile, Q_1: 71.

4. Find the median of the second half of numbers, the third quartile, Q_3: 87.

5. Find the interquartile range (IQR): $IQR = Q_3 - Q_1 = 87 - 71 = 16$.

6. Multiply the IQR by 1.5: $16 \times 1.5 = 24$.

7. Add this number to Q_3 and subtract this number from Q_1.

   $87 + 24 = 111$ and $71 - 24 = 47$

8. Any number smaller than 47 or larger than 111 would be considered an outlier. Therefore, 36 is the only outlier in this set.

1.2 Displaying Data with Graphs

• It is often helpful to display a given data set graphically. Graphing the data of interest can help us use and understand the data more effectively. Make sure you are comfortable creating and interpreting the types of graphs that follow. These include: boxplots, histograms, stemplots, dotplots, bar graphs, and pie charts.

Modified Boxplots

• Modified boxplots are extremely useful in AP Statistics. A modified boxplot is ideal when you are interested in checking a distribution for outliers or skewness, which will be essential in later chapters. To construct a modified boxplot, we use the five-number summary. The box of the modified boxplot consists of Q1, M, and Q3. Outliers are marked as separate points. The tails of the plot consist of either the smallest and largest numbers or the smallest and largest numbers that are not considered outliers by our mathematical criterion discussed earlier. Outliers appear as separate dots or asterisks. Modified boxplots can be constructed with ease using the graphing calculator or computer software. Be sure to use the modified boxplot instead of the regular boxplot, since we are usually interested in knowing if outliers are present. Side-by-side boxplots can be used to make visual comparisons between two or more distributions. Figure 1.4 displays the test scores from Data Set B. Notice that the test score of 36 (which is an outlier) is represented using a separate point.

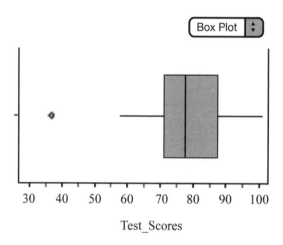

Figure 1.4   Modified boxplot of Data Set B: Test Scores.

Histograms

• Histograms are also useful for displaying distributions when the variable of interest is numeric (Figure 1.5). When the variable is categorical, the graph is called a bar chart or bar graph. The bars of the histogram should be touching and should be of equal width. The heights of the bars represent the frequency or relative frequency. As with modified boxplots, histograms can be easily constructed using the TI-83/84 graphing calculator or computer software. With some minor adjustments to the window of the graphing calculator, we can easily transfer the histogram from calculator to paper. We often use the ZoomStat function of the TI-83/84 graphing calculator to create histograms. ZoomStat will fit the data to the screen of the graphing calculator and often creates bars with non-integer dimensions. In order to create histograms that have integer dimensions, we must make adjustments to the window of the graphing calculator. Once these adjustments have been made, we can then easily copy the calculator histogram onto paper. Histograms are especially useful in finding the shape of a distribution. To find the center of the histogram, as measured by the median, find the line that would divide the histogram into two equal parts. To find the mean of the distributions, locate the balancing point of the histogram.

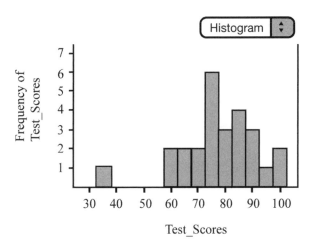

Figure 1.5   Histogram of Data Set B: Test Scores.

Stemplots

• Although we cannot construct a stemplot using the graphing calculator, we can easily construct a stemplot (Figure 1.6) on paper. Stemplots are useful for finding the shape of a distribution as long as there are relatively few data values. Typically, we arrange the data in ascending order. It is often appropriate to round values before graphing. Although our graphing calculators cannot construct a stemplot for us, we can still create a list and order the data in ascending order using the calculator or computer software. Stemplots can have single or "split" stems. Sometimes split stems are used to see the distribution in more detail. Back-to-back stemplots are sometimes used when comparing two distributions. A key should be included with the stemplot so that the reader can interpret the data (i.e: |5|2 = 52.) It is relatively easy to find the five-number summary and describe the distribution once the stemplot is made.

3	6
4	
5	8
6	579
7	133566799
8	15557889
9	38
10	0

Figure 1.6   Stemplot of Data Set B: Test Scores.

Dotplots

• Dotplots can be used to display a distribution (Figure 1.7). Dotplots are easily constructed as long as there are not too many data values. As always, be sure to label and scale your axes and title your graph. Although a dotplot cannot be constructed on the TI-83/84, most statistical software packages can easily construct them.

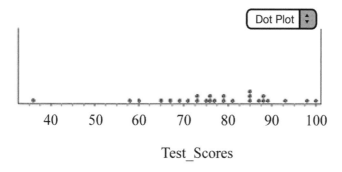

Figure 1.7 Dotplot of Data Set B: Test Scores.

Bar Graphs

• Bar graphs are often used to display categorical data. Bar graphs, unlike histograms, have spaces between the different categories of the variable. The order of the categories is irrelevant and we can use either counts or percentages for the vertical axis. There are only two categories in this bar graph showing soccer goals scored by my two children, Cassidy and Nolan (Figure 1.8). It should be noted that on any given day Cassidy could score more goals than Nolan or vice versa. I had to make one of them have more goals for visual effect only.

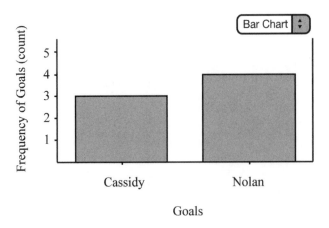

Figure 1.8 Bar graph: Soccer goals made by Cassidy and Nolan.

Pie Charts

• Pie charts are also used to display categorical data (Figure 1.9). Pie charts can help us determine what part of the entire group each category forms. Again, be sure to title your graph and label or code each piece of the pie. On the AP Statistics examination, graphs without appropriate labeling or scaling are considered incomplete.

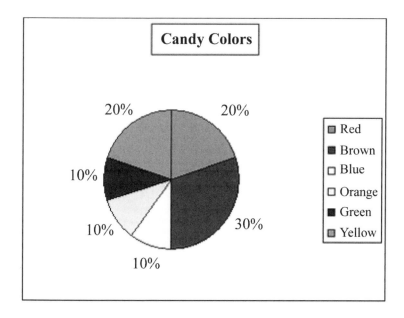

Figure 1.9   Pie chart of candy colors.

Chapter

2

Exploring and Graphing Bivariate Data

2.1 Scatterplots

• Scatterplots are ideal for exploring the relationship between two quantitative variables. When constructing a scatterplot we often deal with explanatory and response variables. The **explanatory variable** may be thought of as the independent variable, and the **response variable** may be thought of as the dependent variable.

• It's important to note that when working with two quantitative variables, we do not always consider one to be the explanatory variable and the other to be the response variable. Sometimes, we just want to explore the relationship between two variables, and it doesn't make sense to declare one variable the explanatory and the other the response.

• We interpret scatterplots in much the same way we interpret univariate data; we look for the overall pattern of the data. We address the **form**, **direction**, and **strength** of the relationship. Remember to look for outliers as well. Are there any points in the scatterplot that deviate from the overall pattern?

• When addressing the **form** of the relationship, look to see if the data is linear (Figure 2.1) or curved (Figure 2.2).

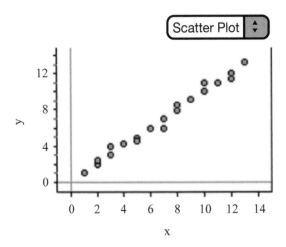

Figure 2.1   Linear relationship.

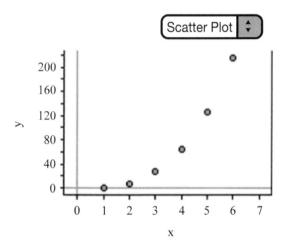

Figure 2.2 Curved relationship.

• When addressing the **direction** of the relationship, look to see if the data has a positive or negative relationship (Figures 2.3, 2.4).

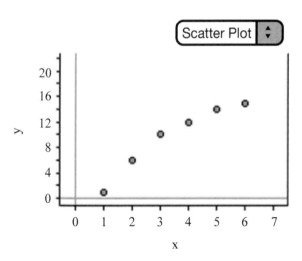

Figure 2.3 Positive, curved relationship.

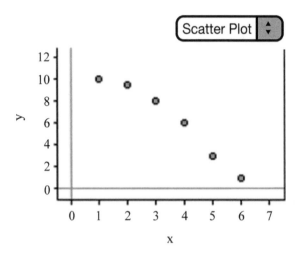

Figure 2.4   Negative, slightly curved relationship.

• When addressing the **strength** of the relationship, consider whether the relationship appears to be weak, moderate, strong, or somewhere in between (Figures 2.5–2.7).

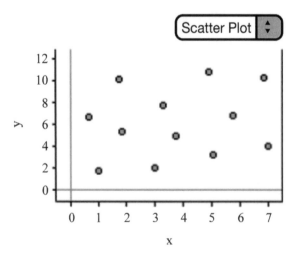

Figure 2.5   Weak or no relationship.

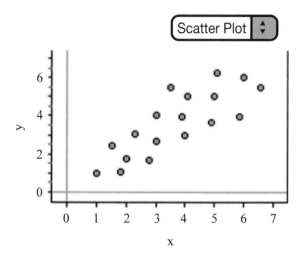

Figure 2.6   Moderate, positive, linear relationship.

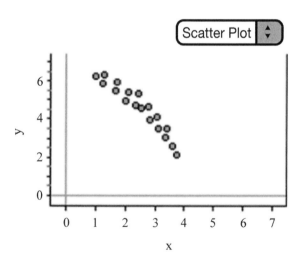

Figure 2.7   Relatively strong, negative, slightly curved relationship.

Correlation

• When dealing with linear relationships, we often use the **r-value,** or the **correlation coefficient.** The correlation coefficient can be found by using the formula:

$$r = \frac{1}{n-1}\sum\left(\frac{x_i - \bar{x}}{s_x}\right)\left(\frac{y_i - \bar{y}}{s_y}\right)$$

• In practice, we avoid using the formula at all cost. However, it helps to suffer through a couple of calculations using the formula in order to understand how the formula works and gain a deeper appreciation of technology.

Facts about Correlation

• It's important to remember the following facts about correlation (make sure you know all of them!):

Correlation (the r-value) only describes a linear relationship. Do not use r to describe a curved relationship.

Correlation makes no distinction between explanatory and response variables. If we switch the x and y variables, we still get the same correlation.

Correlation has no unit of measurement. The formula for correlation uses the means and standard deviations for x and y and thus uses standardized values.

If r is positive, then the association is positive; if r is negative, then the association is negative.

$-1 \leq r \leq 1$: $r = 1$ implies that there is a perfectly linear positive relationship. $r = -1$ implies that there is a perfectly linear negative relationship. $r = 0$ implies that there is no correlation.

The **r-value**, like the mean and standard deviation, is **not a resistant measure**. This means that even one extreme data point can have a dramatic effect on the r-value. Remember that outliers can either strengthen or weaken the r-value. So use caution!

The r-value does not change when you change units of measurement. For example, changing the x and/or y variables from centimeters to millimeters or even from centimeters to inches does not change the r-value.

Correlation does not imply causation. Just because two variables are strongly associated or even correlated (linear) does not mean that changes in one variable are causing changes in another.

Least Squares Regression

• When modeling linear data, we use the **Least Squares Regression Line (LSRL)**. The LSRL is fitted to the data by minimizing the sum of the squared residuals. The graphing calculator again comes to our rescue by calculating the LSRL and its equation. The LSRL equation takes the form of $\hat{y} = a + bx$ where b is the slope and a is the y-intercept. The AP* formula sheet uses the form $\hat{y} = b_0 + b_1 x$. Either form may be used as long as you define your variables. Just remember that the number in front of x is the slope, and the "other" number is the y-intercept.

• Once the LSRL is fitted to the data, we can then use the LSRL equation to make predictions. We can simply substitute a value of x into the equation of the LSRL and obtain the predicted value, $\hat{y}$.

• The LSRL minimizes the sum of the squared residuals. What does this mean? A **residual** is the difference between the observed value, y, and the predicted value, $\hat{y}$. In other words, *residual = observed − predicted*. Remember that all predicted values are located on the LSRL. A residual can be positive, negative, or zero. A residual is zero only when the point is located on the LSRL. Since the sum of the residuals is always zero, we square the vertical distances of the residuals. The LSRL is fitted to the data so that the sum of the square of these vertical distances is as small as possible.

• The slope of the regression line (LSRL) is important. Consider the time required to run the last mile of a marathon in relation to the time required to run the first mile of a marathon. The equation $\hat{y} = 1.25x$, where x is the time required to run the first mile in minutes and $\hat{y}$ is the predicted time it takes to run the last mile in minutes, could be used to model or predict the runner's time for his last mile. The interpretation of the slope in context would be that for every *one* minute increase in time needed to run the first mile, the predicted time to run the last mile would *increase* by 1.25 minutes, on average. It should be noted that the slope is a rate of change and that that since the slope is positive, the time will increase by 1.25 minutes. A negative slope would give a negative rate of change.

Facts about Regression

All LSRLs pass though the point $(\bar{x}, \bar{y})$.

The formula for the slope is $b_1 = r \dfrac{s_y}{s_x}$.

This formula is given on the AP* Exam. Notice that if r is positive, the slope is positive; if r is negative, the slope is negative.

By substituting $(\bar{x}, \bar{y})$ into $\hat{y} = b_0 + b_1 x$ we obtain $\bar{y} = b_0 + b_1\bar{x}$. Solving for b_0, we obtain the y-intercept: $b_0 = \bar{y} - b_1\bar{x}$. This formula is also given on the AP* Exam.

The r^2 value is called the **Coefficient of Determination.** The r^2 value is the proportion of variability of y that can be explained or accounted for by the linear relationship of y on x. To find r^2, we simply square the r-value. **Remember, even an r^2 value of 1 does not necessarily imply any cause-and-effect relationship!** Note: A common misinterpretation of the r^2 value is that it is the percentage of observations (data points) that lie on the LSRL. This is simply not the case. You could have an r^2 value of .70 (70%) and not have any data points that are actually on the LSRL.

It's important to remember the effect that outliers can have on regression. If removing an outlier has a dramatic effect on the slope of the LSRL, then the point is called an **influential observation.** These points have "leverage" and tend to be outliers in the x-direction. Think of prying something open with a pry bar. Applying pressure to the end of the pry bar gives us more leverage or impact. These observations are considered influential because they have a dramatic impact on the LSRL—they pull the LSRL toward them.

• **Example:** Consider the following scatterplot.

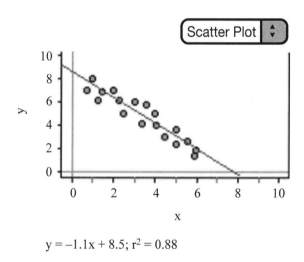

$$y = -1.1x + 8.5; r^2 = 0.88$$

Figure 2.8   Relatively strong, negative, linear relationship.

• We can examine the scatterplot in Figure 2.8 and describe the form, direction, and strength of the relationship. We observe that the relationship is negative—that is, as x increases y decreases. We also note that the relationship is relatively strong and linear. We can write the equation of the LSRL and graph the line. The equation of the LSRL is $\hat{y} = 8.5 - 1.1x$. The correlation coefficient is $r \approx -.9381$ and the coefficient of determination is $r^2 = .88$. Notice that the slope and the r-value are both negative. This is not a coincidence.

• Notice what happens to the LSRL, r, and r^2 as we shift a data point from the scatterplot that is located toward the end of the LSRL in the x-direction. Consider Figure 2.9. The equation of the LSRL changes to $\hat{y} = 7.5 - .677x$, r changes to $\approx .5385$, and r^2 changes to .29. Moving the data point has a dramatic effect on r, r^2, and the LSRL, so we consider it to be an influential observation.

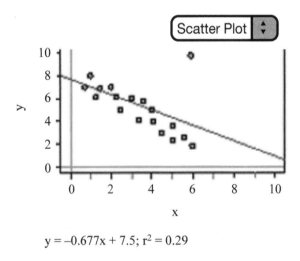

$$y = -0.677x + 7.5; \ r^2 = 0.29$$

Figure 2.9   Influential observation.

• Moving a data point near the middle of the scatterplot does not typically have as much of an impact on the LSRL, r, and r^2 as moving a data point toward the end of the scatterplot in the x-direction. Consider Figure 2.10. Although dragging a data point from the middle of the scatterplot still changes the location and equation of the LSRL, it does not impact the regression nearly as much. Note that r and r^2 change to .7874 and .62, respectively. Although moving this data point impacts regression somewhat, the effect is much less, so we consider this data point "less influential."

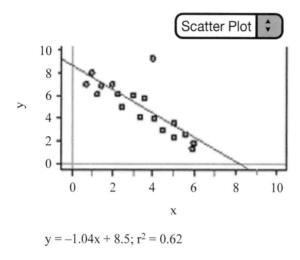

$$y = -1.04x + 8.5; r^2 = 0.62$$

Figure 2.10   Less influential observation.

2.2 Modeling Data

• Linear data can be modeled using the LSRL. It's important to remember, however, that not all data is linear. How do we determine if a line is really the best model to use to represent the data? Maybe the data follow some type of curved relationship?

• Examining the scatterplot, as mentioned earlier, is the first step to finding an appropriate model. However, sometimes looking at the scatterplot and finding the r-value can be a little deceiving. Consider the following two scatterplots. Both contain the same data but are scaled differently. Changing the scale of the scatterplot can make the data appear more or less linear than is really the case. You might guess the r-value of Figure 2.12 to be higher than that of Figure 2.11 since the data points "appear" closer together in Figure 2.12 than they do in Figure 2.11. The r-values are the same, however; only the scale has been changed. Our eyes can sometimes deceive us.

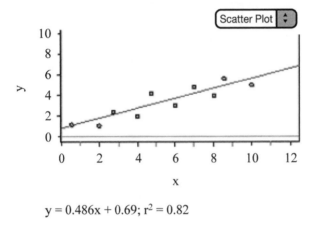

$$y = 0.486x + 0.69; \ r^2 = 0.82$$

Figure 2.11   Figures 2.11 and 2.12 contain the same data.

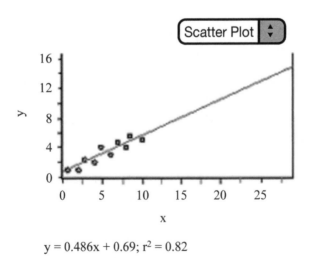

$$y = 0.486x + 0.69; \ r^2 = 0.82$$

Figure 2.12   The r-values are the same in Figures 2.11 and 2.12.

To help make the decision of which model is best, we turn our attention to residual plots.

• The residual plot plots the residuals against the explanatory variable. If the residual plot models the data well, the residuals should not follow a systematic or definite pattern (Figures 2.13–2.14).

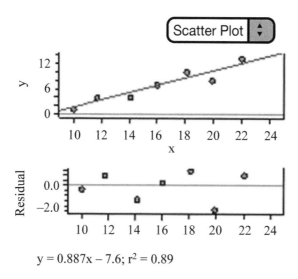

$y = 0.887x - 7.6; r^2 = 0.89$

Figure 2.13   Residual plot with random scatter.

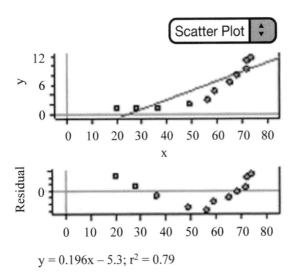

$$y = 0.196x - 5.3; r^2 = 0.79$$

Figure 2.14   Residual plot with a definite pattern.

• The next three examples will be used to aide in the understanding of how to find an appropriate model (equation) for a given data set. The promising AP Stats student (yes, that's *you!*) should understand how to take a given set of bivariate data, determine which model is appropriate, perform the inverse transformation, and write the appropriate equation. The TI-83/84 can be used to construct a scatterplot and the corresponding residual plot. Remember that the graphing calculator will create a list of the residuals once linear regression has been performed on the data. After the appropriate model is determined, we can obtain the LSRL equation from the calculator and transform it to the appropriate equation to model the data. We use logarithms in exponential and power models because these models involve equations with exponents. Remember that a logarithm is just another way to write an exponent. It's important to remember the following algebraic properties of logarithms:

1. $\log(AB) = \log A + \log B$
2. $\log(A|B) = \log A - \log B$
3. $\log X^n = n\log X$

• **Linear Model:** Consider the data in Figure 2.15. Examining the scatterplot of the data reveals a strong, positive, linear relationship. The lack of a pattern in the residual plot confirms that a linear model is appropriate, compared to any other non-linear model. We should be able to get pretty good predictions using the LSRL equation.

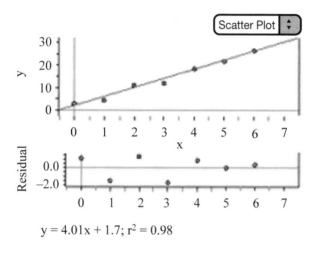

$y = 4.01x + 1.7; r^2 = 0.98$

	x	y
1	0	2.8
2	1	4.1
3	2	11.0
4	3	11.9
5	4	18.5
6	5	21.7
7	6	26.0

Figure 2.15   Scatterplot of linear data and "random" residual plot.

• **Exponential Model:** Consider the data in Figure 2.16. There appears to be a curved pattern to the data. The data does not appear to be linear. To rule out a linear model, we can use our calculator or statistical software to find the LSRL equation and then construct a residual plot of the residuals against *x*. We can see that the residual plot has a definite pattern and thus contradicts a linear model. This implies that a non-linear model is more appropriate.

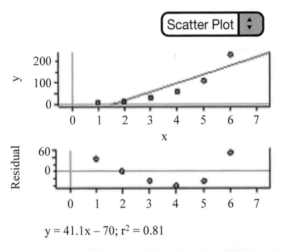

$y = 41.1x - 70; r^2 = 0.81$

	x	y	log_y
=			log(y)
1	1	6.36	0.803457
2	2	13.03	1.11494
3	3	27	1.43136
4	4	55	1.74036
5	5	111	2.04532
6	6	230	2.36173

Figure 2.16   The original data is curved; the residual plot shows a definite pattern.

An exponential or power model might be appropriate. Exponential growth models increase by a fixed percentage of the previous amount. In other words,

$$\frac{230}{111} \approx 2.0721 \quad \frac{111}{55} \approx 2.0182 \quad \frac{55}{27} \approx 2.0370$$

and so on. These percentages are approximately equal. This is an indication that an exponential model might best represent the data. Next, we look at the graph of log y vs. x (Figure 2.17). Notice that the graph of log y vs. x straightens the data. This is another sign that an exponential model might be appropriate. Finally, we can see that the residual plot for the exponential model (log y on x) appears to have random scatter. An exponential model is appropriate.

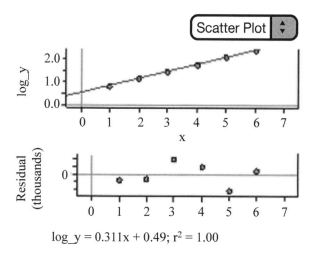

$\log\_y = 0.311x + 0.49;\ r^2 = 1.00$

Figure 2.17   Scatterplot of log y vs. x and residual plot with random scatter.

We can write the LSRL equation for the transformed data of log y vs. x. We then use the properties of logs and perform the inverse transformation as follows to obtain the exponential model for the original data.

1. **Write the LSRL for log y on x.**

   Your calculator may give you $\hat{y} = .4937 + .3112x$, but remember that you are using the log of the y values, so be sure to use $\log \hat{y}$, not just $\hat{y}$.

   $$\log \hat{y} = .4937 + .3112x$$

2. **Rewrite as an exponential equation. Remember that log is the common log with base 10.**

   $$\hat{y} = 10^{.4937 + .3112x}$$

3. **Separate into two powers of 10.**

   $$\hat{y} = 10^{.4937} \cdot 10^{.3112x}$$

4. **Take 10 to the .4937 power and 10 to the .3112 power and rewrite.**

   $$\hat{y} = 3.1167 \cdot 2.0474^{x}$$

Our final equation is an exponential equation. Notice how well the graph of the exponential equation models the original data (Figure 2.18).

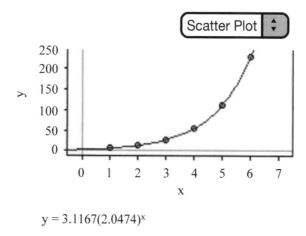

$$y = 3.1167(2.0474)^x$$

Figure 2.18   Exponential equation models the data.

• **Power Model:** Consider the data in Figure 2.19. As always, remember to plot the original data. There appears to be a curved pattern. We can confirm that a linear model is not appropriate by interpreting the residual plot of the residuals against x, once our calculator or software has created the LSRL equation. The residual plot shows a definite pattern; therefore a linear model is not appropriate.

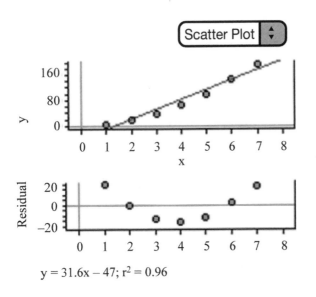

y = 31.6x − 47; r² = 0.96

	x	y	log_x	log_y
=			log(x)	log(y)
1	1	3.9	0	0.591065
2	2	16.2	0.30103	1.20952
3	3	35.5	0.477121	1.55023
4	4	64.5	0.60206	1.80956
5	5	99.6	0.69897	1.99826
6	6	145.0	0.778151	2.16137
7	7	192.0	0.845098	2.2833

Figure 2.19   The original data is curved; the residual plot shows a definite pattern.

We can then examine the graph of log y on x. Notice that taking the log of the y-values and plotting them against x does not straighten the data— in fact, it bends the data in the opposite direction. What about the residual plot for log y vs. x? There appears to be a pattern in the residual plot (Figure 2.20); this indicates that an exponential model is not appropriate.

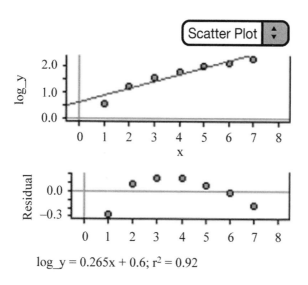

log_y = 0.265x + 0.6; r^2 = 0.92

Figure 2.20   Scatterplot of log y vs. x. The residual plot shows a definite pattern.

Next, we plot log y vs. log x (Figure 2.21). Notice that this straightens the data and that the residual plot of log y vs. log x appears to have random scatter. A power model is therefore appropriate.

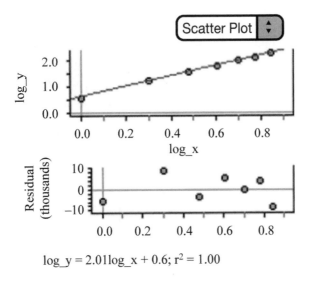

$$\log\_y = 2.01\log\_x + 0.6; \; r^2 = 1.00$$

Figure 2.21   Scatterplot of log y vs. log x. The residual plot
                  shows random scatter.

We can then perform the inverse transformation to obtain the appropriate
equation to model the data.

1. **Write the LSRL for log y on log x.**

   $\log \hat{y} = .5970 + 2.0052 \log x$ Remember we are using logs!

2. **Rewrite as a power equation.**

   $\hat{y} = 10^{.5970+2.0052\log x}$

3. **Separate into two powers of 10.**

   $\hat{y} = 10^{.5970} \cdot 10^{2.0052\log x}$

4. Use the power property of logs to rewrite.

$$\hat{y} = 10^{.5970} \cdot 10^{\log x^{2.0052}}$$

5. Take 10 to the .5970 power and cancel 10 to the log power.

$$\hat{y} = 3.9537 \cdot x^{2.0052}$$

Our final equation is a power equation. Notice how well the power model fits the data in the original scatterplot (Figure 2.22).

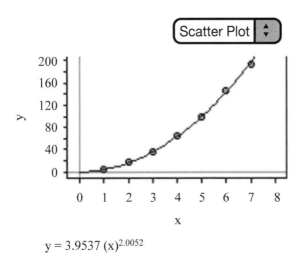

$$y = 3.9537\,(x)^{2.0052}$$

Figure 2.22   The power model fits the data.

Chapter

3

Normal Distributions

3.1 Density Curves

• Density curves are smooth curves that can be used to describe the overall pattern of a distribution. Although density curves can come in many different shapes, they all have something in common: **The area under any density curve is always equal to one.** This is an extremely important concept that we will utilize in this and other chapters. It is usually easier to work with a smooth density curve than a histogram, so we sometimes overlay the density curve onto the histogram to approximate the distribution. A specific type of density curve called a *normal curve* will be addressed in section 3.2. This "bell-shaped" curve is especially useful in many applications of statistics as you will see later on. We describe density curves in much the same way we describe distributions when using graphs such as histograms or stemplots.

• The relationship between the mean and the median is an important concept, especially when dealing with density curves. In a symmetrical density curve, the mean and median will be equal if the distribution is perfectly symmetrical or approximately equal if the distribution is approximately symmetrical. If a distribution is skewed left, then the mean will be "pulled" in the direction of the skewness and will be less than the median. If a distribution is skewed right, the mean is again "pulled" in the direction of the skewness and will be greater than the median. Figure 3.1 displays distributions that are skewed left, skewed right, and symmetrical. Notice how the mean is "pulled" in the direction of the skewness.

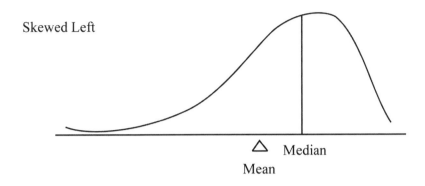

Skewed Left

△ Median
Mean

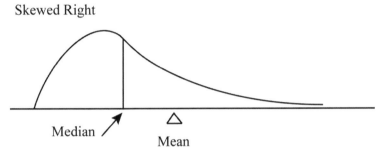

Skewed Right

Median ↗
Mean

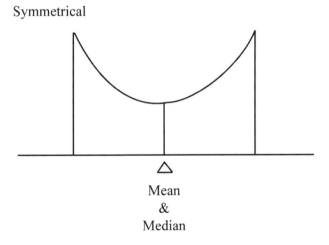

Symmetrical

△
Mean
&
Median

Figure 3.1 The relationship of mean and median in skewed and symmetrical distributions.

• It's important to remember that the mean is the "balancing point" of the density curve or histogram and that the median divides the density curve or histogram into two parts, equal in area (Figure 3.2).

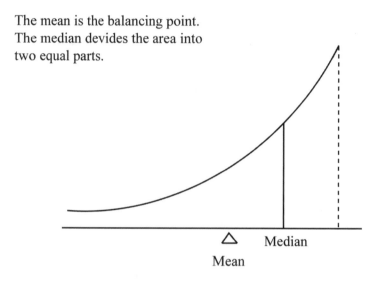

The mean is the balancing point.
The median devides the area into
two equal parts.

Median

Mean

Figure 3.2   The mean is the "balancing point" of the distribution. The median divides the density curve into two equal areas.

3.2 Normal Distributions

• One particular type of density curve that is especially useful in statistics is the normal curve, or **normal distribution.** Although all normal distributions have the same overall shape, they do differ somewhat depending on the mean and standard deviation of the distribution (Figure 3.3). If we increase or decrease the mean while keeping the standard deviation the same, we will simply shift the distribution to the right or to the left. The more we increase the standard deviation, the "wider" and "shorter" the density curve will be. If we decrease the standard deviation, the density curve will be "narrower" and "taller." Remember that all density curves, including normal curves, have an area under the curve equal to one. So, no matter what value the mean and standard deviation take, the area under the normal curve is equal to one. This is very important, as you'll soon see.

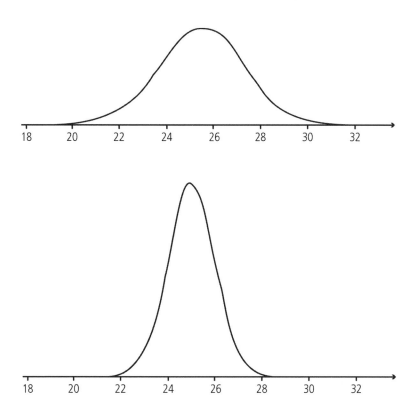

Figure 3.3   Two normal distributions with different standard deviations.

The equation for the standard normal curve is: $y = \dfrac{1}{\sqrt{2\pi}} e^{-x^2/2}$

The Empirical Rule (the 68, 95, 99.7 Rule)

• All normal distributions follow the Empirical Rule. That is to say that all normal distributions have: 68% of the observations falling within σ (one standard deviation) of the mean, 95% of the observations falling within 2σ (two standard deviations) of the mean, and 99.7% (almost all) of the observations falling within 3σ (three standard deviations) of the mean (Figure 3.4).

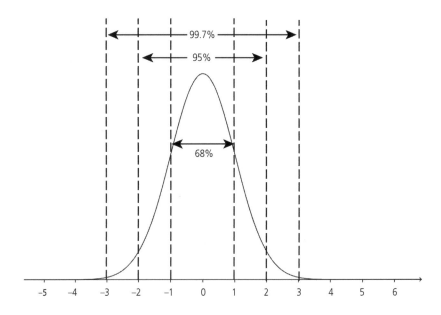

Figure 3.4   About 68% of observations fall within one standard deviation, 95% within two standard deviations, and 99.7% within three standard deviations.

• **Example 1:** Let's assume that the number of miles that a particular tire will last roughly follows a normal distribution with $\mu = 40,000$ miles and $\sigma = 5000$ miles. Note that we can use shorthand notation $N(40,000, 5000)$ to denote a normal distribution with mean equal to 40,000 and standard deviation equal to 5,000. Since the distribution is not exactly normal but approximately normal, we can assume the distribution will roughly follow the 68, 95, 99.7 Rule. Using the 68, 95, 99.7 Rule we can conclude the following (see Figure 3.5):

About 68% of all tires should last between 35,000 and 45,000 miles ($\mu \pm \sigma$)

About 95% of all tires should last between 30,000 and 50,000 miles ($\mu \pm 2\sigma$)

About 99.7% of all tires should last between 25,000 and 55,000 miles ($\mu \pm 3\sigma$)

Using the 68, 95, 99.7 Rule a little more creatively, we can also conclude:

About 34% of all tires should last between 40,000 and 45,000 miles.

About 34% of all tires should last between 35,000 and 40,000 miles.

About 2½ % of all tires should last more than 50,000 miles.

About 84% of all tires should last less than 45,000 miles.

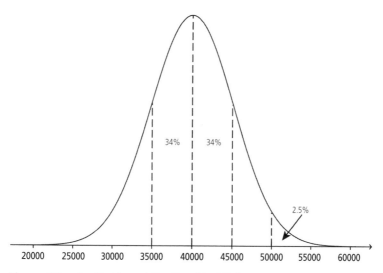

Figure 3.5 Application of the Empirical Rule.

3.3 Normal Calculations

• **Example 2:** Referring back to Example 1, let's suppose that we want to determine the percentage of tires that will last more than 53,400 miles. Recall that we were given $N(40{,}000, 5000)$. To get a more exact answer than we could obtain using the Empirical Rule, we can do the following:

Solution: Always make a sketch! (See Figure 3.6.)

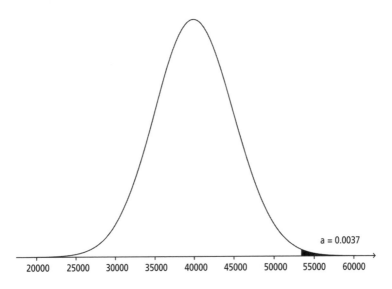

Figure 3.6   Make a sketch and shade to the right of 53,400.

Shade the area that you are trying to find, and label the mean in the center of the distribution. Remember that the mean and median are equal in a normal distribution since the normal curve is symmetrical.

Obtain a standardized value (called a **z-score**) using $z = \dfrac{x - \mu}{\sigma}$.

Using substitution, we obtain $z = \dfrac{53,400 - 40,000}{5000} = 2.68$

Notice that the formula for z takes the difference of x and μ and divides it by σ. Thus, a z-score is the number of standard deviations that x lies above or below the mean. So, 53,400 is 2.68 standard deviations above the mean. You should always get a positive value for z if the value of x is above the mean, and a negative value for z if the value of x is below the mean.

When we find the z-score, we are standardizing the values of the distribution. Since these values are values of a normal distribution, the distribution we obtain is called the **standard normal distribution.** This new distribution, the standard normal distribution, has a mean of zero and a standard deviation of one. We can then write $N(0,1)$

The advantage of standardizing any given normal distribution to the standard normal distribution is that we can now find the area under the curve for any given value of x that is needed.

We can now use the z-score of 2.68 that we obtained earlier. Using Table A, we can look up the area to the left of $z = 2.68$. Notice that Table A has two sides—one for positive values for z and the other for negative values for z. Using the side of the table with the positive values for z, follow the left-hand column down until you reach 2.6. Then go across the top of the table until you reach .08. By cross-referencing 2.6 and .08, we can obtain the area to the left of $z = 2.68$, which is 0.9963.

In other words, 99.63% of tires will last less than 53,400 miles. We want to know what percent of tires will last *more than* 53,400 miles, so we subtract 0.9963 from 1. Remember that the total area under any density curve is equal to one.

We obtain $1 - 0.9963 = 0.0037$.

Conclude in context.

That is, only 0.37% of tires will last more than 53,400 miles. We can also state that the probability that a randomly chosen tire of this type will last longer than 53,400 miles is equal to 0.0037.

• **Example 3:**   Again referring to Example 1, find the probability that a randomly chosen tire will last between 32,100 miles and 41,900 miles.

Make a sketch. (See Figure 3.7.)

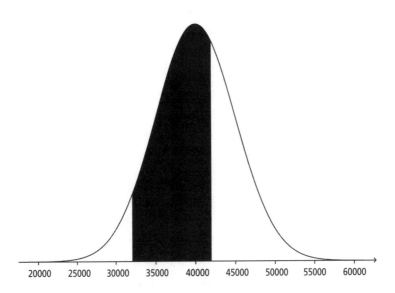

Figure 3.7   Make a sketch and shade between 32,100 and 41,900.

Locate the mean on the normal curve as well as the values of 32,100 and 41,900. Shade the area between 32,100 and 41,900.

Calculate the z-scores.

$$z = \frac{32,100 - 40,000}{5000} = -1.58 \text{ and } z = \frac{41,900 - 40,000}{5000} = 0.38$$

Find the areas to the left of −1.58 *and* 0.38 using Table A.

The area to the left of −1.58 is equal to 0.0571, and the area to the left of 0.38 is equal to 0.6480.

Since we want to know the probability that a tire will last between 32,100 and 41,900 miles, we will subtract the two areas. **Remember that any area that we look up in Table A is the area to the left of** z**.**

$$0.6480 - 0.0571 = 0.5909$$

Conclude in context:

The probability that a randomly chosen tire will last between 32,100 miles and 41,900 miles is equal to 0.5909.

• **Example 4:** Consider a national mathematics exam where the distribution of test scores roughly follows a normal distribution with mean, $\mu = 320$, and standard deviation, $\mu = 32$. What score must a student obtain to be in the top 10% of all students taking the exam?

Make a sketch! (See Figure 3.8.)

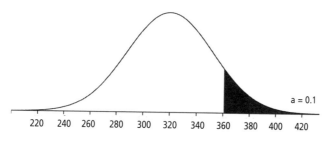

Figure 3.8 Make a sketch and shade the top 10%.

Shade the appropriate area.

Use the formula for z.

$$z = \frac{x - \mu}{\sigma}$$

Using substitution, we obtain:

$$z = \frac{x - 320}{32}$$

In order to solve for x, we need to obtain an appropriate value of z. Using Table A "backwards," we look in the body of the table for the value closest to 0.90, which is 0.8997. The value of z that corresponds to an area to the left of 0.8997 is 1.28, so $z = 1.28$. Again, remember that everything we look up in Table A is the area to the *left* of z, so we look up what's closest to 0.90, not 0.10.

Substituting for z, we obtain:

$$1.28 = \frac{x - 320}{32}$$

Solving for x, we obtain:

$$x = 360.96$$

Conclude in context.

A student must obtain a score of approximately 361 in order to be in the top 10% of all students taking the exam.

• **Example 5:** Consider the national mathematics test in Example 4. The middle 90% of students would score between which two scores?

Make a sketch! (See Figure 3.9.)

Shade the appropriate area.

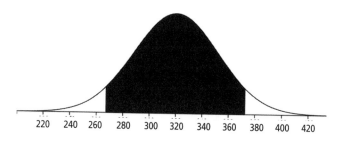

Figure 3.9   Make a sketch and shade the middle 90%.

Use the formula for z.

$$z = \frac{x - \mu}{\sigma}$$

Using substitution, we obtain:

$$z = \frac{x - 320}{32}$$

In order to solve for x, we need to obtain an appropriate value of z. Consider that we are looking for the middle 90% of test scores. Remembering once again that the area under the normal curve is 1, we can obtain the area on the "outside" of 90%, which would be 10%. This forms two "tails," which we consider the right and left tails.

These "tails" are equal in area and thus have an area of 0.05 each. We can then use Table A, as we did in Example 4, to obtain a z-score that corresponds to an area of 0.05. Notice that two values are equidistant from 0.05. These areas are 0.0495 and 0.0505, which correspond to z-scores of -1.64 and -1.65, respectively. Since the areas we are looking up are the same distance away from 0.05, we split the difference and go out one more decimal place for z. We use $z = -1.645$.

$$-1.645 = \frac{x - 320}{32}$$

Solving for x, we obtain:

$$x = 267.36$$

We can now find the test score that would be the cutoff value for the top 5% of scores. Notice that since the two tails have the same area, we can use $z = 1.645$. The z-scores are opposites due to the symmetry of the normal distribution.

$$1.645 = \frac{x - 320}{32}$$

Solving for x, we obtain:

$$x = 372.64$$

Conclude in context.

The middle 90% of students will obtain test scores that range from approximately 267 to 373.

Assessing Normality

• Inferential statistics is a major component of the AP Statistics curriculum. When you infer something about a population based on sample data, it is often important to assess the normality of a population. We can do this by looking at the number of observations in the sample that lie within one, two, and three standard deviations from the mean. In other words, use the Empirical Rule. Do approximately 68, 95, and 99.7% of the observations fall within $\mu \pm 1\sigma$, $\mu \pm 2\sigma$, and $\mu \pm 3\sigma$? Larger data sets should roughly follow the 68, 95, 99.7 Rule while smaller data sets typically have more variability and therefore may be less likely to follow the Empirical Rule despite coming from normal populations.

• We can also look at a graph of the sample data. By constructing a histogram, stemplot, modified boxplot, or line plot, we can examine the data to look for strong skewness and outliers. Non-normal populations often produce sample data that have skewness or outliers or both. Normal populations are more likely to have sample data that are symmetrical and bell-shaped and usually do not have outliers.

• **Normal probability plots** can also be used to assess the normality of a population through sample data. A normal probability plot is a scatterplot that graphs a predicted z-score against the value of the variable. Most graphing calculators and statistical software packages are capable of constructing normal probability plots. You should be much more concerned with how to interpret a normal probability plot than with how one is constructed. Again, technology helps us out in constructing the plot.

• Interpret the normal probability plot by assessing the linearity of
the plot. The more linear the plot, the more normal the distribution.
A non-linear probability plot is a good sign of a non-normal population.
Consider the following data taken from a distribution known to be
uniform and non-normal (Figure 3.10). The accompanying normal
probability plot is curved and is thus a sign that the data is indeed taken
from a non-normal population.

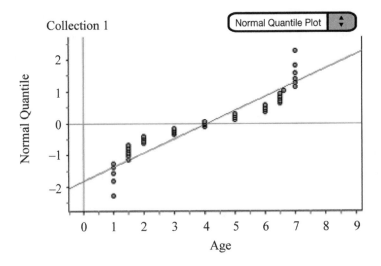

Normal Quantile = 0.443age − 1.8

Figure 3.10   The non-linearity of the normal probability plot suggests that
the data comes from a non-normal population.

Chapter

4

Samples, Experiments, and Simulations

• It is imperative that we follow proper data collection methods when gathering data. **Statistical inference** is the process by which we draw conclusions about an entire population based on sample data. Whether we are designing an experiment or sampling part of a population, it's critical that we understand how to correctly gather the data we use. Improper data collection leads to incorrect assumptions and predictions about the population of interest. If you learn nothing else about statistics, I hope you learn to be skeptical about how data is collected and to interpret the data correctly. Properly collected data can be extremely useful in many aspects of everyday life. Inference based on data that was poorly collected or obtained can be misleading and lead us to incorrect conclusions about the population.

4.1 Sampling

• You will encounter certain types of sampling in AP Statistics. As always, it's important that you fully understand all the concepts discussed in this chapter. We begin with some basic definitions.

• A **population** is all the individuals in a particular group of interest. We might be interested in how the student body of our high school views a new policy about cell phone usage in school. The population of interest is all students in the school. We might take a poll of some students at lunch or during English class on a particular day. The students we poll are considered a **sample** of the entire population. If we sample the entire student body, we are actually conducting a census. A **census** consists of all individuals in the entire population. The U.S. Census attempts to count every resident in the United States and is required by the Constitution every ten years. The data collected by the U.S. Census will help determine the number of seats each state has in the House of Representatives. There has even been some political debate on whether or not the U.S. should spend money trying to count everyone when information could be gained by using appropriate sampling techniques.

• A **sampling frame** is a list of individuals from the entire population from which the sample is drawn.

• Several different types of sampling are discussed in AP Statistics. One type often referred to is an **SRS, or simple random sample.** An SRS is a sample in which every set of *n* individuals has an equal chance of being chosen. Referring back to our population of students, we could conduct an SRS of size 100 from the 2200 students by numbering all students from 1 to 2200. We could then use the random integer function on our calculator, or the table of random digits, or we could simply draw 100 numbers out of a hat that included the 2200 numbers. It's important to note that, in an SRS, not only does every individual in the population have an equal opportunity of being chosen, but so does each sample.

• When using the table of random digits, we should remember a few things. First of all, there are many different ways to use the table. We'll discuss one method and then I'll briefly give a couple examples of another way that the table might be used. Consider our population of 2200 students. After numbering each student from 0001 to 2200, we can go to the table of random digits (found in many statistics books). We can go to any line—let's say line 145. We can look at the first four-digit number, which is 1968. This would be the first student selected for our sample. The next number is 7126. Since we do not have a student numbered 7126, we simply skip over 7126. We also skip 3357, 8579, and 5806. The next student chosen is 0993. We continue in this fashion until we've selected the number of students we want in our sample. If we get to the end of the line, we simply go on the next line. This is only one method we could use. A different method might be to start at the top with line 101. Use the first "chunk" of five digits and use the last four digits of that five-digit number (note that the numbers are grouped in groups of five for the purpose of making the table easier to read). That would give us 9223, which we would skip. We could then either go across to the next "chunk" of five digits or go down to the "chunk" of five digits below our first group. No matter how you use the random digit table, just remember to be consistent and stay with the same system until the entire sample has been chosen. Skipping the student numbered 1559 because he's your old boyfriend or she's your old girlfriend is not what random sampling is all about.

• A **stratified random sample** could also be used to sample our student body of 2200 students. We might break up our population into groups that we believe are similar in some fashion. Maybe we feel that freshmen, sophomores, juniors, and seniors will feel different about our new policy concerning cell phones. We call these homogeneous groups **strata**. Within each stratum, we would then conduct an SRS. We would then combine these SRSs to obtain the total sample. Stratified random sampling guarantees representation from each strata. In other words, we know that our sample includes the opinions of freshmen, sophomores, juniors, and seniors.

• A **cluster sample** is similar to a stratified sample. In a cluster sample, however, the groups are heterogeneous, not homogeneous. That is, we don't feel like the groups will necessarily differ from one another. Once the groups are determined, we can conduct an SRS within each group and form the entire sample from the results of each SRS. Usually this method is used to make the sampling cheaper or easier. We might sample our 2200 students during our three lunch periods. We could form three SRSs from our three lunch groups as long as we feel that all three lunch groups are similar to one another and all represent the population equally.

- **Systematic sampling** is a method in which it is predetermined how the sample will be obtained. We might, for example, sample every 25th student from our list of 2200 students. We should note that this method is not considered an SRS since not all samples of a given size have an equal chance of being chosen. Think about it this way: If we sample every 25th student of the 2200, that's 88 students. The first 25 students on the list of the 2200 students would never be chosen together, so technically it's not an SRS.

- A **convenience sample** could also be conducted from our 2200 students. We would conduct a convenience sample because it's, well, convenient. We might sample students in the commons area near the cafeteria because it's an easy thing to do and we can do so during our lunch break. It should be noted, however, that convenience samples almost always contain **bias.** That is to say that they tend to systematically understate or overstate the proportion of people that feel a certain way; they are usually not representative of the entire population.

- A **voluntary response sample** could be obtained by having people respond on their own. We might try to sample some of our 2200 students by setting up an online survey where students could respond one time to a survey if they so choose. These types of samples suffer from **voluntary response bias** because those that feel very strongly either for or against something are much more willing to respond. Those that feel strongly against something are actually more likely to respond than those that have strong positive feelings.

- A **multistage sample** might also be used. This is sampling that combines several different types of sampling. Some national opinion polls are conducted using this method.

• We should also be concerned with how survey questions are worded. We should ensure that the wording is not slanted in such a way as to sway the person taking the survey to answer the question in a particular manner. Poorly worded questions can lead to **response bias.** Training sometimes takes place so that the person conducting the survey interview uses good interviewing techniques.

• **Undercoverage** occurs when individuals in the population are excluded in the process of choosing the sample. Undercoverage can lead to bias, so caution must be used.

• **Nonresponse** can also lead to bias when certain selected individuals cannot be reached or choose not to participate in the sample.

• Our goal is to eliminate bias. Through proper sampling, it is possible to eliminate a good deal of the bias that can be present if proper sampling is not used. We must realize that sampling is never perfect. If I draw a sample from a given population and then draw another sample in the exact same manner, I rarely get the exact same results. There is almost always some **sampling variability**. Think about sampling our student body of 2200 students. If we conduct an SRS of 25 students and then conduct another SRS of 25 students, we will probably not be sampling the same 25 students and thus may not get the exact same results. More discussion about sampling viability will take place in later chapters. Remember, however, that larger random samples will give more accurate results than smaller samples conducted in the same manner. A smaller random sample, however, may give more accurate results than a larger non-random sample.

4.2 Designing Experiments

• Now that we've discussed some different types of sampling, it's time to turn our attention to experimental design. It's important to understand both observational studies and experiments and the difference between them. In an **observational study,** we are observing individuals. We are studying some variable about the individuals but not imposing any treatment on them. We are simply studying what is already happening. In an **experiment,** we are actually imposing a treatment on the individuals and studying some variable associated with that treatment. The treatment is what is applied to the subjects or experimental units. We use the term "subjects" if the experimental units are humans. The treatments may have one or more **factors,** and each factor may have one or more **levels**.

• **Example 1:** Consider an experiment where we want to test the effects of a new laundry detergent. We might consider two factors: water temperature and laundry detergent. The first factor, temperature, might have three levels: cold, warm, and hot water. The second factor, detergent, might have two levels: new detergent and old detergent. We can combine these to form six treatments as listed in Figure 4.1.

Temperature

		Cold	Warm	Hot
Type of Detergent	New	New/Cold	New/Warm	New/Hot
	Old	Old/Cold	Old/Warm	Old/Hot

Figure 4.1   Six treatments.

• It's important to note that we cannot prove or even imply a cause-and-effect relationship with an observational study. We can, however, prove a cause-and-effect relationship with an experiment. In an experiment, we observe the relationship between the explanatory and response variables and try to determine if a cause-and-effect relationship really does exist.

• The first type of experiment that we will discuss is a completely randomized experiment. In a **completely randomized experiment**, subjects or experimental units are randomly assigned to a treatment group. Completely randomized experiments can be used to compare any number of treatments. Groups of equal size should be used, if possible.

• **Example 2:** Consider an experiment in which we wish to determine the effectiveness of a new type of arthritis medication. We might choose a completely randomized design. Given 600 subjects suffering from arthritis, we could randomly assign 200 subjects to group 1, which would receive the new arthritis medication, 200 subjects to group 2, which would receive the old arthritis medication, and 200 subjects to group 3, which would receive a **placebo,** or "dummy" pill. To ensure that the subjects were randomly placed into one of the three treatment groups, we could assign each of the 600 subjects a number from 001 to 600. Using the random integer function on our calculator, we could place the first 200 subjects whose numbers come up in group 1, the second 200 chosen in group 2, and the remaining subjects in group 3. It should be noted that a placebo is used to help control the **placebo effect,** which comes into play when people respond to the "idea" that they are receiving some type of treatment. A placebo, or "dummy" pill, is used to ensure that the placebo effect contributes equally to all three groups. The placebo should taste, feel, and look like the real medication. The subjects would take the medication for a predetermined period of time before the effectiveness of the medication was evaluated. We can use a diagram to help outline the design (see Figure 4.2).

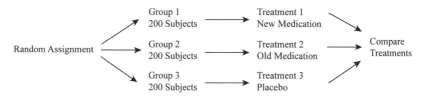

Figure 4.2   A completely randomized design.

To describe an experiment, it can be useful (but not essential) to use a diagram. Remember to explain how you plan to randomly assign individuals to each treatment in the experiment. This can be as simple as using the table of random digits or using the random integer function of the graphing calculator. Be specific in your diagram, and be sure to fully explain how you are setting up the experiment.

• **Example 3:** Let's reconsider Example 2. Suppose there is reason to believe that the new arthritis medication might be more effective for men than for women. We would then use a type of design called a **block design.** We would divide our group of 600 subjects into one group of males and one group of females. Once our groups were blocked on gender, we would then randomly assign our group of males to one of the three treatment groups and our group of females to one of the three treatment groups. It's important to note that the use of blocking reduces variability within each of the blocks. That is, it eliminates a confounding variable that may systematically skew the results. For example, if one is conducting an experiment on a weight-loss pill and blocking is not used, the random assignment of the subjects may assign more females to the experimental group. If males and females respond differently to the treatment, you will not be able to determine whether the weight loss is due to the drug's effectiveness or due to the gender of the subjects in the group. Be sure to include random assignment in your diagram, but make sure that you've done so after you've separated males and females. There are often a few students who get in a hurry on an exam and randomly place subjects into groups of males and females. It's good that they remember that random assignment is important, but it needs to come after the blocking, not before.

• The arthritis experiment in Examples 2 and 3 might be either a **single-blind** or **double-blind experiment.** In a single-blind experiment, the person taking the medication would not know whether they had the new medication, the old medication, or the placebo. If a physician is used to help assess the effectiveness of the treatments, the experiment should probably be double-blind. That is, neither the subject receiving the treatment nor the physician would know which treatment the subject had been given. Obviously, in the case of a double-blind experiment, there must be a third-party member that knows which subjects received the various treatments.

• **Example 4:** A manufacturer of bicycle tires wants to test the durability of a new material used in bicycle tires. A completely randomized design might be used where one group of cyclists uses tires made with the old material and another group uses tires made with the new material. The manufacturer realizes that not all cyclists will ride their bikes on the same type of terrain and in the same conditions. To help control for these variables, we can implement a **matched-pairs design.** Matching is a form of blocking. One way to do this is to have each cyclist use both types of tires. A coin toss could determine whether the cyclist uses the tire with the new material on the front of the bike or on the rear. We could then compare the front and rear tire for each cyclist. Another way to match in this situation might be to pair up cyclists according to rider size and weight, the location where they ride, and/or the type of terrain they typically ride on. A coin could then be tossed to decide which of the two cyclists uses the tires with the new material and which uses the tires with the old material. This method might not be as effective as having each cyclist serve as his/her own control and use one tire of each type.

• When you're designing various types of experiments, it's important to remember the *four principles of experimental design.* They are:

1. **Control.** It is very important to control the effects of confounding variables. **Confounding variables** are variables (aside from the explanatory variable) that may affect the response variable. We often use a control group to help assess whether or not a particular treatment actually has some effect on the subjects or experimental units. A control group might receive the "old" (or "traditional") treatment, or it might receive a placebo ("dummy" pill). This can help compare the various treatments and allow us to determine if the new treatment really does work or have a desired effect.

2. **Randomization.** It's critical to reduce bias (systematic favoritism) in an experiment by controlling the effects of confounding variables. We hope to spread out the effects of these confounding variables by using chance to randomly assign subjects or experimental units to the various treatments.

3. **Replication.** There are two forms of replication that we must consider. First, we should always use more than one or two subjects or experimental units to help reduce chance variation in the results. The more subjects or experimental units we use, the better. By increasing the number of experimental units or subjects, we know that the difference between the experimental group and the control group is really due to the imposed treatment(s) and not just due to chance. Second, we should have designed an experiment that can be replicated by others doing similar research.

4. **Blocking.** Blocking is not a requirement for experimental design,
   but it may help improve the design of the experiment in some cases.
   Blocking places individuals who are similar in some characteristic
   in the same group, or "block." These individuals are expected to
   respond in a similar manner to the treatment being imposed. For
   example, we may have reason to believe that men and women will
   differ in how they are affected by a particular type of medication.
   In this case, we would be blocking on gender. We would form one
   group of males and one group of females. We would then use randomization to assign males and females to the various treatments.

4.3 Simulation

• Simulation can be used in statistics to model random or chance behavior.
In much the same way an airplane simulator models how an actual aircraft
flies, simulation can be used to help us predict the probability of some
real-life occurrences. For our purposes in AP Statistics, we'll try to keep
it simple. If you are asked to set up a simulation in class or even on an
exam, keep it simple. Use things like the table of random digits, a coin, a
die, or a deck of cards to model the behavior of the random phenomenon.

• Let's set up an example: As I was walking out of the grocery store a
few years ago, my two children, Cassidy and Nolan (ages 5 and 7 at the
time), noticed a lottery machine that sold "scratch-offs" near the exit of
the store. Despite explaining to them how the "scratch-offs" worked and
that the probability of winning was, well … not so good, they persuaded
me to partake in the purchase of three $1 "scratch-offs." Being an AP
Stats teacher and all, I knew I had a golden opportunity to teach them a
lesson in probability and a "lesson" that gambling was "risky business."

Sure, we might win a buck or two, but chances were pretty good that we'd lose, and even if we did win, the kids would hopefully lose interest since we would most likely just be getting our money back. Once we were in the car, the lesson began. "Hmmm …"Odds are 1:4," I told them. That means that on average, you win about one time for every five times you play. I carefully explained that the chances of winning were not very good and that if we won, chances were pretty good that we would not win a lot. Two "scratch-offs" later … two winners, $1 each. Hmmm…. Not exactly what I had planned, but at least I had my $2 back. "Can we buy some more?" they quickly asked. I told them that the next time we stopped for gas, we could buy two more "scratch-offs" but that was it. Surely they'd learn their lesson this time. Two weeks later, we purchased two more $1 "scratch-offs." Since I was in a hurry, I handed each of them a coin and a "scratch-off" and away we drove. Unfortunately for Nolan, his $1 "scratch off" resulted in a loss. I felt a little bad about his losing, but in the long run it would probably be best. Moments later, Cassidy yells out, "I won a hundred dollars!" Sure, I thought. She's probably just joking. "Let me see that!" I quickly pulled over at the next opportunity to realize that she had indeed won $100! Again, not exactly what I'd planned, but hey … it was $100! What are the chances of winning on three out of four "scratch-offs"? Let's set up a simulation to try to answer the question.

• **Example 5:** Use simulation to find the probability that someone who purchases four $1 "scratch-offs" will win something on three out of the four "scratch-offs." Assume the odds of winning on the "scratch-off" are 1:4.

Solution: If the odds of winning are 1:4, that means that in the long run we should expect to win one time out of every five plays. That is, we should expect, out of five plays, to win once and lose four times, on average. In other words, the probability of winning is 1/5. Sometimes we might win more than expected and sometimes we might win less than expected, but we should average one win for every four losses. We can set up a simulation to estimate the probability of winning.

Let the digits 0–1 represent a winning "scratch-off."

Let the digits 2–9 represent a losing "scratch-off."

Note that 0–1 is actually 2 numbers and 2–9 is 8 numbers. Also note that 1:4 odds would be the same as 2:8 odds. We have used single-digit numbers for the assignment as it is the simplest method in this case. We could have used double-digit numbers for the assignment, but this would be unnecessary. Several different methods would work as long as the odds reduce to 1:4. To make it easier to keep track of the numbers, we will group the one-digit numbers in "chunks" of four and label each group "W" for win and "L" for lose. Each group of four one-digit numbers represents one simulation of purchasing four "scratch-offs" (Figure 4.3). Starting at line 107 of the table of random digits, we obtain:

L	L	L	L	L	L	L
8273	9578	9020	8074	7511	8167	6553
L	L	L	L	L	L	L
6094	0720	2417	8682	4943	6179	0906
L	L	L	L	L	L	L
3600	9193	6515	4123	9638	8545	3468
L	L	L	L	L	L	L
3844	8487	8918	3382	4697	3936	4420
L	L	L	L	L	W	L
8148	6694	8760	5130	9297	0041	2712

Figure 4.3   "Scratch-off" probability simulation.

Figure 4.3 displays 50 trials of purchasing four "scratch-off" tickets. Only one of the 50 trials produced three winning tickets out of four. Based on our simulation, the probability of winning three out of four times is only 1/50. In other words, Cassidy and Nolan were pretty lucky. Students sometimes find simulation to be a little tricky. Remember to keep it as simple as possible. A simulation does not need to be complicated to be effective.

Chapter

5

Probability

5.1 **Probability and Probability Rules**

• An understanding of the concept of randomness is essential for tackling the concept of probability. What does it mean for something to be random? AP Statistics students usually have a fairly good concept of what it means for something to be random and have likely done some probability calculations in their previous math courses. I'm always a little surprised, however, when we use the random integer function of the graphing calculator when randomly assigning students to their seats or assigning students to do homework problems on the board. It's almost as if students expect everyone in the class to be chosen before they are chosen for the second or third time. Occasionally, a student's number will come up two or even three times before someone else's, and students will comment that the random integer function on the calculator is not random. Granted, it's unlikely for this to happen with 28 students in the class, but not impossible. Think about rolling a standard six-sided die. The outcomes associated with this event are **random**—that is, they are uncertain but follow a predictable distribution over the long run. The proportion associated with rolling any one of the six sides of the die over the long run is the **probability** of that outcome.

• It's important to understand what is meant by *in the long run*. When I assign students to their seats or use the random integer function of the graphing calculator to assign students to put problems on the board, we are experiencing what is happening *in the short run*. **The Law of Large Numbers** tells us that the long-run relative frequency of repeated, independent trials gets closer to the expected relative frequency once the number of trials increases. Events that seem unpredictable in the short run will eventually "settle down" after enough trials are accumulated. This may require many, many trials. The number of trials that it takes depends on the variability of the random variable of interest. The more variability, the more trials it takes. Casinos and insurance companies use the Law of Large Numbers on an everyday basis. Averaging our results over many, many individuals produces predictable results. Casinos are guaranteed to make a profit because they are in it for the long run whereas the gambler is in it for the relative short run.

• The **probability** of an event is always a number between 0 and 1, inclusive. Sometimes we consider the **theoretical probability** and other times we consider the **empirical probability**. Consider the experiment of flipping a fair coin. The theoretical probability of the coin landing on either heads or tails is equal to 0.5. If we actually flip the coin a 100 times and it lands on tails 40 times, then the empirical probability is equal to 0.4. If the empirical probability is drastically different from the theoretical probability, we might consider whether the coin is really fair. Again, we would want to perform many, many trials before we conclude that the coin is unfair.

• **Example 1:** Consider the experiment of flipping a fair coin three times. Each flip of the coin is considered a **trial** and each trial for this experiment has two possible **outcomes,** heads or tails. A list containing all possible outcomes of the experiment is called a **sample space.** An **event** is a subset of a sample space. A **tree diagram** can be used to organize the outcomes of the experiment, as shown in Figure 5.1.

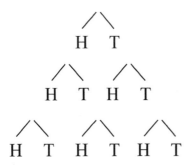

Figure 5.1   Tree diagram.

Tree diagrams can be useful in some problems that deal with probability. Each trial consists of one line in the tree diagram, and each branch of the tree diagram can be labeled with the appropriate probability. Working our way down and across the tree diagram, we can obtain the eight possible outcomes in the sample space. $S = \{HHH, HHT, HTH, HTT, THH, THT, TTH, TTT\}$ To ensure that we have the correct number of outcomes listed in the sample space, we could use the **counting principle,** or **multiplication principle.** The multiplication principle states that if you can do task 1 in m ways and you can do task 2 in n ways, then you can do task 1 followed by task 2 in $m \times n$ ways. In this experiment we have three trials, each with two possible outcomes. Thus, we would have $2 \times 2 \times 2 = 8$ possible outcomes in the sample space.

• **Example 2:** Let's continue with the experiment discussed in Example 1. What's the probability of flipping the coin three times and obtaining heads all three times?

Solution: We can answer that question in one of two ways. First, we could use the sample space. *HHH* is one of eight possible (equally likely) outcomes listed in the sample space, so $P(HHH) = \frac{1}{8}$. The second method we could use to obtain $P(HHH)$ is to use the concept of **independent events.** Two events are **independent** if the occurrence or non-occurrence of one event does not alter the probability of the second event. The trials of flipping a coin are independent. Whether or not the first flip results in heads or tails does not change the probability of the coin landing on heads or tails for the second or third flip. If two events are independent, then $P(A \cap B) = P(A \text{ and } B) = P(A) \cdot P(B)$. We can apply this concept to this experiment. $P(HHH) = (\frac{1}{2}) \cdot (\frac{1}{2}) \cdot (\frac{1}{2}) = \frac{1}{8}$. Later on, in Example 11, we will show how to prove whether or not two events are independent.

• **Example 3:** Again consider Example 1. Find the probability of obtaining at least one tail (not all heads).

Solution: The events "all heads" and "at least one tail" are **complements**. The set "at least one tail" is the set of all outcomes from the sample space excluding "all heads." All the outcomes in a given sample space should sum to one, and so any two events that are complements should sum to one as well. Thus, the probability of obtaining "at least one tail" is equal to: $1 - P(HHH) = 1 - \frac{1}{8} = \frac{7}{8}$. We can verify our answer by examining the sample space we obtained in Example 1 and noting that 7 out of the 8 equally likely events in the sample space contain at least one tail. Typical symbols for the complement of event A are: A^c, A', or $\bar{A}$.

• **Example 4:** Consider the experiment of drawing two cards from a standard deck of 52 playing cards. Find the probability of drawing two hearts if the first card is replaced and the deck is shuffled before the second card is drawn. The following tree diagram can be used to help answer the question.

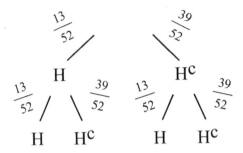

Figure 5.2   Tree diagram with replacement.

Solution: Let *H* = "*heart*" *and Hc* = "*non – heart*" Notice that the probability of the second card being a heart is independent of the first card being a heart. Thus, $P(HH) = \frac{1}{4} \cdot \frac{1}{4} = \frac{1}{16}$.

• **Example 5:** How would Example 4 change if the first card were not replaced before the second card was drawn? Find the probability of drawing two hearts if the first card drawn is not replaced before the second card is drawn. Notice how the probabilities in the tree diagram change depending on whether or not a heart is drawn as the first card.

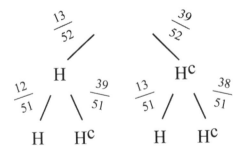

Figure 5.3   Tree diagram without replacement.

When two events *A and B* are not independent, then $P(A \cap B) =$
$P(A) \cdot P(B \mid A)$ This is a **conditional probability,** which we will
discuss in more detail in section 5.2. Applying this formula, we obtain
$P(HHH) = {}^{13}\!\!/_{52} \cdot {}^{12}\!\!/_{51} = {}^{1}\!\!/_{17}$.

• **Example 6:** Suppose that in a particular high school the probability
that a student takes AP Statistics is equal to 0.30 (call this event A), and
the probability that a student takes AP Calculus is equal to 0.45 (call this
event B.) Suppose also that the probability that a student takes both AP
Statistics and AP Calculus is equal to 0.10. Find the probability that a
student takes either AP Statistics or AP Calculus.

Solution: We can organize the information given in a Venn diagram as
shown in Figure 5.4.

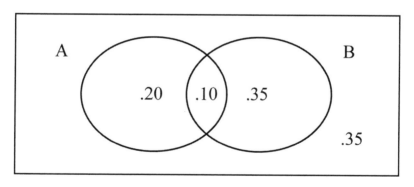

Figure 5.4   Venn diagram for two events, A and B.

Notice the probability for each section of the Venn diagram. The total circle for event A (AP Statistics) has probabilities that sum to 0.30 and the total circle for event B (AP Calculus) has probabilities that sum to 0.45. Also notice that all four probabilities in the Venn diagram sum to 1. We can use the **General Addition Rule for the Union of Two Events.** $P(A \cup B) = P(A) + P(B) - P(B) - P(A \cap B)$ Note that $\cup$ (union) means "or" and $\cap$ (intersection) means "and." We could then apply the formula as follows:

$$P(A \cup B) = 0.30 + 0.45 - 0.10 = 0.65.$$

If you consider the Venn diagram, the General Addition Rule makes sense. When you consider event A, you are adding in the "overlapping" of the two circles (student takes AP Stats and AP Calculus), and when you consider event B, you are again adding in the "overlapping" of the two circles. Thus, the General Addition Rule has us subtracting the intersection of the two circles, which is the "overlapping" section.

• **Example 7:** Reconsider Example 6. Find the probability that a student takes neither AP Statistics nor AP Calculus.

Solution: From the Venn diagram in Figure 5.4 we can see that the probability that a student takes neither course is the area (probability) on the outside of the circles, which is 0.35. We could also conclude that 20% of students take AP Statistics but not AP Calculus and that 35% of students take AP Calculus but not AP Statistics.

• **Example 8:** Referring again to Example 6, suppose that AP Statistics and AP Calculus are taught only once per day and during the same period. It would then be impossible for a student to take both AP Statistics and AP Calculus. How would the Venn diagram change?

Solution: We could construct the Venn diagram as shown in Figure 5.5.

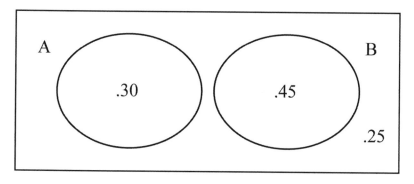

Figure 5.5   Venn diagram for disjoint (mutually exclusive) events.

Notice that the circles are not overlapping since events A and B cannot occur at the same time. Events *A and B* are **disjoint,** or **mutually exclusive.** This implies that $P(A \cap B) = 0$. Applying the General Addition Rule, we obtain: $P(A \cup B) = 0.30 + 0.45 = 0.75$. Notice that we do not have to subtract $P(A \cap B)$ since it is equal to 0. Thus, for **disjoint events**:

$$P(A \cup B) = P(A) + P(B).$$

• Don't confuse independent events and disjoint (mutually exclusive) events. Try to keep these concepts separate, but remember that if you know two events are independent, they cannot be disjoint. The reverse is also true. If two events are disjoint, then they cannot be independent. Think about Example 8, where it was impossible to take AP Statistics and AP Calculus at the same time (disjoint events.) If a student takes AP Statistics, then the probability that they take AP Calculus changes from 0.35 to zero. Thus, these two events, which are disjoint, are not independent. That is, taking AP Statistics changes the probability of taking AP Calculus. It's also worth noting that some events are neither disjoint nor independent. The fact that an event is not independent does not necessarily mean it's disjoint and vice versa. Consider drawing two cards at random, without replacement, from a standard deck of 52 playing cards. The events "first card is an ace" and "second card is an ace" are neither disjoint nor independent. The events are not independent because the probability of the second card being an ace depends on whether or not an ace was drawn as the first card. The events are not disjoint because it is possible that the first card is an ace and the second card is also an ace.

5.2 Conditional Probability and Bayes's Rule

• **Example 9:** Example 5 is a good example of what we mean by **conditional probability.** That is, finding a given probability if it is known that another event or condition has occurred or not occurred. Knowing whether or not a heart was chosen as the first card determines the probability that the second card is a heart. We can find P(*2nd card heart* | *1st card heart*) by using the formula given in Example 5 and solving for P(A | B), read *A given B*.

Thus, $P(A / B) = \dfrac{P(A \cap B)}{P(B)}$.

When applying the formula, just remember that the numerator is always the intersection ("and") of the events, and the denominator is always the event that comes after the "given that" line. Applying the formula, we obtain:

$$P(2nd\ card\ heart\ /\ 1st\ card\ heart) = \frac{P(2nd\ card\ heart \cap 1st\ card\ heart)}{P(1st\ card\ heart)} = \frac{\frac{12}{51} \cdot \frac{13}{52}}{\frac{13}{52}} = \frac{12}{51}$$

The formula works, although we could have just looked at the tree diagram and avoided using the formula. Sometimes we can determine a conditional probability simply by using a tree diagram or looking at the data, if it's given. The next problem is a good example of a problem where the formula for conditional probability really comes in handy.

• **Example 10:** Suppose that a medical test can be used to determine if a patient has a particular disease. Many medical tests are not 100% accurate. Suppose the test gives a positive result 90% of the time if the person really has the disease and also gives a positive result 1% of the time when a person does not have the disease. Suppose that 2% of a given population actually have the disease. Find the probability that a randomly chosen person from this population tests positive for the disease.

Solution: We can use a tree diagram to help us solve the problem (Figure 5.6).

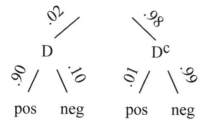

Figure 5.6   Tree diagram for conditional probability.

It's important to note that the test can be positive whether or not the person actually has the disease. We must consider both cases. Let event $D = $ *Person has the disease* and let event $D^c = $ *Person does not have the disease*. Let "*pos*" $= positive$ and "*neg*" $= negative$.

$$P(pos) = 0.02 \cdot 0.90 + 0.98 \cdot 0.01 = 0.0278$$

Thus, the probability that a randomly chosen person tests positive for the disease is 0.0278.

• **Example 11:** Referring to Example 10, find the probability that a randomly chosen person has the disease given that the person tested positive. In this case we know that the person tested positive and we are trying to find the probability that they actually have the disease. This is a conditional probability known as **Bayes's Rule.**

$$P(D / pos) = \frac{P(D \cap pos)}{P(pos)} = \frac{0.02 \cdot 0.90}{0.02 \cdot 0.90 + 0.98 \cdot 0.01} \approx 0.6475.$$

You should understand that Bayes's Rule is really just an extended conditional probability rule. However, it's probably unnecessary for you to remember the formula for Bayes's Rule. If you understand how to apply the conditional probability formula and you can set up a tree diagram, you should be able to solve problems involving Bayes's Rule. Just remember that to find $P(pos)$ you have to consider that a positive test result can occur if the person has the disease and if the person does not have the disease.

• Again, in some problems you may be given the probabilities and need to use the conditional probability formula and in others it may be unnecessary. In the following example, the formula for conditional probability certainly works, but using it is unnecessary. All the data needed to answer the conditional probability is given in the table (Figure 5.7).

• **Example 12:** Consider a marathon in which 38,500 runners participate. Figure 5.7 contains the times of the runners broken down by age. Find the probability that a randomly chosen runner runs under 3 hours given that they are in the 50+ age group.

	18–29	30–49	50+	Total
Under 3 Hrs.	451	527	19	997
3 – Under 4	3,280	4,215	1,518	9,013
4 – Under 5	5,167	10,630	3,563	19,360
Over 5 Hrs.	4,219	3,879	1,032	9,130
Total	13,117	19,251	6,132	38,500

Figure 5.7   Age and time of runners in a marathon.

Solution: Although we could use the conditional probability formula, it's really unnecessary. It's given that the person chosen is in the 50+ age group, which means instead of dividing by the total number of runners (38,500) we can simply divide the number of runners that are both 50+ and ran under 3 hours by the total number of 50+ runners.

$$P(Under\ 3\ hrs\ /\ 50+) = \frac{19}{6132}$$

Thus, the probability that a randomly chosen person runs under 3 hours for the marathon given that they are 50 or older is about 0.0031.

• **Example 13:** Let's revisit independent events. Is the age of a runner independent of the time that the runner finishes the marathon? It doesn't seem likely that the two events are independent of one another. We would expect older runners to run slower, on average, than younger runners. There are certainly exceptions to this rule, which I am reminded of when someone ten years older than me finishes before me in a marathon! Are the events "finishes under 3 hours" and "50+ age group" independent of one another?

Solution: Remember, if the two events are independent then:

$$P(A \cap B) = P(A) \cdot P(B)$$

Thus, $P(under\ 3\ hrs\ and\ 50+) = P(under\ 3\ hrs) \cdot P(50+)$

We can again use the table values from Figure 5.7 to answer this question.

$$\frac{19}{38,500} = \frac{997}{38,500} \cdot \frac{6132}{38,500}$$

$$0.0004935 \neq 0.0041245$$

Since the two are not equal, the events are not independent.

5.3 Discrete Random Variables

• Now that we've discussed the concepts of randomness and probability, we turn our attention to random variables. A **random variable** is a numeric variable from a random experiment that can take on different values. The random variable can be discrete or continuous. A **discrete random variable, X,** is a random variable that can take on only a countable number. (In some cases a discrete random variable can take on a finite number of values and in others it can take on an infinite number of values.) For example, if I roll a standard six-sided die, there are only six possible values of X, which can take on the values 1, 2, 3, 4, 5, or 6. I can then create a valid probability distribution for X, which lists the values of X and the corresponding probability that X will occur (Figure 5.8).

x	1	2	3	4	5	6
P(x)	1/6	1/6	1/6	1/6	1/6	1/6

Figure 5.8 Probability distribution.

The probabilities in a valid probability distribution must all be between 0 and 1 and all probabilities must sum to 1.

• **Example 14:** Consider the experiment of rolling a standard (fair) six-sided die and the probability distribution in Figure 5.8. Find the probability of rolling an odd number greater than 1.

Solution: Remember that this is a discrete random variable. This means that rolling an odd number greater than 1 is really rolling a 3 or a 5. Also note that we can't roll a 3 and a 5 with one roll of the die, which makes the events disjoint or mutually exclusive. We can simply add the probabilities of rolling a 3 and a 5.

$$P(3 \text{ or } 5) = \tfrac{1}{6} + \tfrac{1}{6} = \tfrac{1}{3}$$

• We sometimes need to find the mean and variance of a discrete random variable. We can accomplish this by using the following formulas:

Mean $\quad \mu_x = x_1 p_1 + x_2 p_2 + ... + x_n p_n$ or $\sum x \cdot P(x)$

Variance $\quad \sigma_x^2 = (x_1 - \mu_x)^2 p_1 + (x_2 - \mu_x)^2 p_2 + ... + (x_n - \mu_x)^2 p_n$ or
$$\sigma_x^2 = \sum (x - \mu_x)^2 \cdot P(x)$$

Std. Dev $\quad \sigma_x = \sqrt{Var(X)}$

Recall that the standard deviation is the square root of the variance, so once we've found the variance it is easy to find the standard deviation. It's important to understand how the formulas work. Remember that the mean is the center of the distribution. The mean is calculated by summing up the product of all values that the variable can take on and their respective probabilities. The more likely a given value of X, the more that value of X is "weighted" when we calculate the mean. The variance is calculated by averaging the squared deviations for each value of X from the mean.

• **Example 15:** Again consider rolling a standard six-sided die and the probability distribution in Figure 5.8. Find the mean, variance, and standard deviation for this experiment.

Solution: We can apply the formulas for the mean and variance as follows:

$$\mu_x = 1(\tfrac{1}{6}) + 2(\tfrac{1}{6}) + \dots + 6(\tfrac{1}{6}) = 3.5$$

$$\sigma_x^2 = (1-3.5)^2(\tfrac{1}{6}) + (2-3.5)^2(\tfrac{1}{6}) + \dots + (6-3.5)^2(\tfrac{1}{6}) \approx 2.9167$$

$$\sigma_x \approx 1.7078$$

Notice that since the six sides of the die are equally likely, it seems logical that the mean of this discrete random variable is equal to 3.5. As always, it's important to show your work when applying the appropriate formulas. Note that you can also utilize the graphing calculator to find the mean, variance, and standard deviation of a discrete random variable. But be careful! Some calculators give the standard deviation, not the variance. That's not a problem, however; if you know the standard deviation, you can simply square it to get the variance. You can find the standard deviation of a discrete random variable on the TI83/84 graphing calculator by creating list one to be the values that the discrete random variable takes on and list two to be their respective probabilities. You can then use the one-variable stats option on your calculator to find the standard deviation. Caution! You must specify that you want one-variable stats for list one and list two (1-Var Stats L_1, L_2). Otherwise your calculator will only perform one-variable stats on list one.

• **Example 16:** Suppose a six-sided die with sides numbered 1–6 is loaded in such a way that in the long run you would expect to have twice as many "1's" and twice as many "2's" as any other outcome. Find the probability distribution for this experiment, and then find the mean and standard deviation.

Solution: See Figure 5.9.

x	1	2	3	4	5	6
P(x)	2x	2x	x	x	x	x

Figure 5.9 Probability distribution for Example 16.

Since we are dealing with a valid probability distribution, we know that all probabilities must sum to 1.

$2x + 2x + x + x + x + x = 8x$

$8x = 1$

$x = \frac{1}{8}$

We can then complete the probability distribution as follows (Figure 5.10).

x	1	2	3	4	5	6
P(x)	2/8	2/8	1/8	1/8	1/8	1/8

Figure 5.10 Probability distribution for Example 16.

We can then find the mean and standard deviation by applying the formulas and using our calculators.

$$\mu_x = 1(\tfrac{1}{4}) + 2(\tfrac{1}{4}) + \ldots + 6(\tfrac{1}{8}) = 3$$

$$\sigma_x^2 = (1-3)^2(\tfrac{1}{4}) + (2-3)^2(\tfrac{1}{4}) + \ldots + (6-3)^2(\tfrac{1}{8}) \approx 3$$

$$\sigma_x \approx 1.7321$$

Notice that the mean is no longer 3.5. The loaded die "weights" two sides of the die so that they occur more frequently, which lowers the mean from 3.5 on the standard die to 3 on the loaded die.

5.4 Continuous Random Variables

• Some random variables are not discrete—that is, they do not always take on values that are countable numbers. The amount of time that it takes to type a five-page paper, the time it takes to run the 100 meter dash, and the amount of liquid that can travel through a drainage pipe are all examples of continuous random variables.

• A **continuous random variable** is a random variable that can take on values that comprise an interval of real numbers. When dealing with probability distributions for continuous random variables we often use density curves to model the distributions. Remember that any density curve has area under the curve equal to one. The probability for a given event is the area under the curve for the range of values of X that make up the event. Since the probability for a continuous random variable is modeled by the area under the curve, the probability of X being one specific value is equal to zero. The event being modeled must be for a range of values, not just one value of X. Think about it this way: The area for one specific value of X would be a line and a line has area equal to zero. This is an important distinction between discrete and continuous random variables. Finding $P(X \geq 3)$ and $P(X > 3)$ would produce the same result if we were dealing with a continuous random variable since $P(X = 3) = 0$. Finding $P(X \geq 3)$ and $P(X > 3)$ would probably produce different results if we were dealing with a discrete random variable. In this case, $X > 3$ would begin with 4 because 4 is the first countable number greater than 3. $X \geq 3$ would include 3.

• It is sometimes necessary to perform basic operations on random variables. Suppose that X is a random variable of interest. The expected value (mean) of X would be μ_x and the variance would be σ_x^2. Suppose also that a new random variable Z can be defined such that $Z = a \pm bx$. The mean and variance of Z can be found by applying the following **Rules for Means and Variances**:

$$\mu_x = a \pm b\mu_x$$

$$\sigma_z^2 = b^2 \sigma_x^2$$

$$\sigma_z = b\sigma_x$$

• **Example 17:** Given a random variable X with $\mu_x = 4$ *and* $\sigma_x = 1.2$, find μ_z, σ_z^2 *and* σ_z given that $Z = 3 + 4X$.

Solution: Instead of going back to all values of X and multiplying all values by 4 and adding 3, we can simply use the mean, variance, and standard deviation of X and apply the Rules for Means and Variances.

Think about it. If all values of X were multiplied by 4 and added to 3, the mean would change in the same fashion. We can simply take the mean of X, multiply it by 4, and then add 3.

$$\mu_z = 3 + 4(4) = 19$$

The variability (around the mean) would be increased by multiplying the values of X by 4. However, adding 3 to all the values of X would increase the values of X by 3 but would not change the variability of the values around the new mean. Adding 3 does not change the variability, so the Rules for Variances does not have us add 3, but rather just multiply by 4 *or* 4^2 depending on whether we are working with the standard deviation or variance. If we are finding the new standard deviation, we multiply by 4; if we are finding the new variance, we multiply by 4^2 *or* 16. When dealing with the variance we multiply by the factor squared. This is due to the relationship between the standard deviation and variance. Remember that the variance is the square of the standard deviation.

$$\sigma_z^2 = 4^2(1.44) = 23.04$$

$$\sigma_z = 4(1.2) = 4.8$$

Notice that $4.8^2 = 23.04$.

- Sometimes we wish to find the sum or difference of two random variables. If X and Y are random variables, we can use the following to find the mean of the sum or difference:

$$\mu_{X+Y} = \mu_X + \mu_Y$$

$$\mu_{X-Y} = \mu_X - \mu_Y$$

We can also find the variance by using the following *if* X and Y are **independent random variables.**

$$\sigma^2_{X+Y} = \sigma^2_X + \sigma^2_Y$$ **This is not a typo! We always add variances!**

$$\sigma^2_{X-Y} = \sigma^2_X + \sigma^2_Y$$

If X and Y are *not* independent random variables, then we must take into account the correlation ρ. It is enough for AP* Statistics to simply know that the variables must be independent in order to add the variances. You do not have to worry about what to do if they are not independent; just know that they have to be independent to use these formulas.

- To help you remember the relationships for means and variances of random variables, consider the following statement that I use in class: "We can add or subtract means, but we only add variances. We never, ever, ever, ever, never, ever add standard deviations. We only add variances." This incorrect use of the English language should help you remember how to work with random variables.

- **Example 18:** John and Gerry work on a watermelon farm. Assume that the average (expected) weight of a Crimson watermelon is 30 lbs. with a standard deviation of 3 lbs. Also assume that the average weight of a particular type of seedless watermelon is 25 lbs. with a standard deviation of 2 lbs. Gerry and John each reach into a crate of watermelons and randomly pull out one watermelon. Find the average weight, variance, and standard deviation of two watermelons selected at random if John picks out a Crimson watermelon and Gerry picks out one of the seedless watermelons.

Solution: The average weight of the two watermelons is just the sum of the two means.

$$\mu_{X+Y} = 30 + 25 = 55$$

To find the variance for each type of melon, we must first square the standard deviation of each type to obtain the variance. We then add the variances.

$$\sigma^2_{X+Y} = 3^2 + 2^2 = 13$$

We can then simply take the square root to obtain the standard deviation.

$$\sigma_{X+Y} \approx 3.6056 \; lbs.$$

Thus, the combined weight of the two watermelons will have an expected (average) weight of 55 lbs. with a standard deviation of approximately 3.6056 lbs.

• **Example 19:** Consider a 32 oz. soft drink that is sold in stores. Suppose that the amount of soft drink actually contained in the bottle is normally distributed with a mean of 32.2 oz. and a standard deviation of 0.8 oz. Find the probability that two of these 32 oz soft drinks chosen at random will have a mean difference that is greater than 1 oz.

Solution: The sum or difference for two random variables that are normally distributed will also have a normal distribution. We can therefore use our formulas for the sum or difference of independent random variables and our knowledge of normal distributions.

Let $X =$ the number of oz. of soft drink in one 32 oz. bottle. We are trying to find:

$$P(\mu_{x_1} - \mu_{x_2}) > 1$$

We know that the mean of the differences is the difference of the means:

$$\mu_x - \mu_{x_2} = 0$$

We can find the standard deviation of the difference by adding the variances and taking the square root.

$$\text{var}(X_1 - X_2) = \sigma^2_{x_1} + \sigma^2_{x_2} = 0.64 + 0.64 = 1.28$$

$$Std\ Dev\ (X_1 - X_2) \approx 1.1314$$

We can now use the mean and standard deviation along with a normal curve to obtain the following:

$$P\left(z > \frac{1-0}{1.1314}\right) = 0.88$$

Using Table A and subtracting from 1 we obtain:

$$1 - 0.8106 = 0.1894$$

The probability that two randomly selected bottles have a mean difference of more than 1 oz. is equal to 0.1894.

5.5 Binomial Distributions

• One type of discrete probability distribution that is of importance is the **binomial distribution.** Four conditions must be met in order for a distribution to be considered a binomial. These conditions are:

1. Each observation can be considered a "success" or "failure." Although we use the words "success" and "failure," the observation might not be what we consider to be a success in a real-life situation. We are simply categorizing our observations into two categories.
2. There must be a fixed number of trials or observations.
3. The observations must be independent.
4. The probability of success, which we call p, is the same from one trial to the next.

• It's important to note that many probability distributions do not fit a binomial setting, so it's important that we can recognize when a distribution meets the four conditions of a binomial and when it does not. If a distribution meets the four conditions, we can use the shorthand notation, $B(n, p)$, to represent a binomial distribution with n trials and probability of success equal to p. We sometimes call a binomial setting a **Bernoulli trial.** Once we have decided that a particular distribution is a binomial distribution, we can then apply the Binomial Probability Model. The formula for a binomial distribution is given on the AP* Statistics formula sheet.

- $P(X = k) = \binom{n}{k} p^k (1-p)^{n-k}$ *where:*

    $n = number\ of\ trials$

    $p = probability\ of\ "success"$

    $1-p = probability\ of\ "failure"$

    $k = number\ of\ successes\ in\ n\ trials$

    $\binom{n}{k} = \dfrac{n!}{(n-k)!k!}$

- **Example 20:** Consider Tess, a basketball player who consistently makes 70% of her free throws. Find the probability that Tess makes exactly 5 free throws in a game where she attempts 10 free throws. (We must make the *assumption* that the free throw shots are independent of one another.)

Solution: We can use the formula as follows:

$\binom{10}{5}(.70)^5(.30)^5$ There are 10 trials and we want exactly 5 trials to

be a we want exactly 5 trials to be a "success." $\binom{10}{5}$ means we have

a combination of 10 things taken 5 at a time in any order. Some textbooks write this as $_{10}C_5$. We can use our calculator to determine that there are in fact 252 ways to take 5 things from 10 things, if we do not care about the order in which they are taken. For example: Tess could make the first 5 shots and miss the rest. Or, she could make the first shot, miss the next 5, and then make the last 4. Or, she could make every other shot of the 10 shots. The list goes on. There are 252 ways she could make exactly 5 of the ten shots.

Notice that the probabilities of success and failure must add up to one since there are only two possible outcomes that can occur. Also notice that the exponents add up to 10. This is because we have 10 total trials with 5 successes and 5 failures.

We should be able to use our calculator to obtain:

$$\binom{10}{5}(.70)^5(.30)^5 \approx .1029.$$

This is the work we would want to show on the AP* Exam!

We could also use the following calculator command on the TI 83/84: binompdf(10,.7,5). We enter 10 for the number of trials, .7 for the probability of success, and 5 as the number of trials we are going to obtain.

However, **binompdf(10,.7,5) does not count as work on the AP* Exam.** You must show $\binom{10}{5}(.70)^5(.30)^5$ even though you might not actually use it to get the answer. Binompdf is a calculator command specific to one type of calculator, not standard statistical notation. Don't think you'll get credit for writing down how to do something on the calculator. You will not! **You must show the formula *or* identify the variable as a binomial as well as the parameters n and p.**

• **Example 21:** Consider the basketball player in Example 20. What is the probability that Tess makes at most 2 free throws in 10 attempts?

Solution: Consider that "at most 2" means Tess can make either 0, or 1, or 2 of her free throws. We write the following:

$$\binom{10}{0}(.70)^0(.30)^{10} + \binom{10}{1}(.70)^1(.30)^9 + \binom{10}{2}(.70)^2(.30)^8$$

Always show this work.

We can either calculate the answer using the formula or use: binomcdf(10,.7,2) Notice that we are using cdf instead of pdf. The "c" in cdf means that we are calculating the cumulative probability. The graphing calculator always starts at 0 trials and goes up to the last number in the command. **Again, the work you show should be the work you write when applying the formula, not the calculator command!**

Either way we use our calculator we obtain:

$$\binom{10}{0}(.70)^0(.30)^{10} + \binom{10}{1}(.70)^1(.30)^9 + \binom{10}{2}(.70)^2(.30)^8 \approx 0.0016$$

• **Example 22:** Again consider Example 20. What is the probability that Tess makes more than 2 of her free throws in 10 attempts?

Solution: Making more than 2 of her free throws would mean making 3 or more of the 10 shots. That's a lot of combinations to consider and write down.

It's easier to use the idea of the complement that we studied earlier in the chapter. Remember that if Tess shoots 10 free throws she could make anywhere between none and all 10 of her shots. Using the concept of the complement we can write:

$$1-\left[\binom{10}{0}(.70)^0(.30)^{10}+\binom{10}{1}(.70)^1(.30)^9+\binom{10}{2}(.70)^2(.30)^8\right]\approx.9984$$

That's pretty sweet. Try to remember this concept. It can make life a little easier for you!

• **Example 23:** Find the expected number of shots that Tess will make and the standard deviation.

The following formulas are given on the AP* Exam:

$$\mu=np \quad and \quad \sigma=\sqrt{np(1-p)}$$

Remember that the expected number is the average number of shots that Tess will make out of every 10 shots.

$$\mu=10\cdot0.7=7$$

Thus, Tess will make 7 out of every 10 shots, on average. Seems logical!

Using the formula for standard deviation, we obtain:

$$\sigma = \sqrt{10(0.7)(0.3)} \approx 1.4491$$

5.6 Geometric Distributions

• **Example 24:** Consider Julia, a basketball player who consistently makes 70% of her free throws. What is the probability that Julia makes her first free throw on her third attempt?

• How does this example differ from that of the previous section? In this example there are not a set number of trials. Julia will keep attempting free throws until she makes one. This is the major difference between binomial distributions and **geometric distributions.**

• There are four conditions that must be met in order for a distribution to fit a geometric setting. These conditions are:

1. Each observation can be considered a "success" or "failure."

2. The observations must be independent.

3. The probability of success, which we call p, is the same from one trial to the next.

4. The variable that we are interested in is the number of observations it takes to obtain the first success.

- The probability that the first success is obtained in the nth observation is: $P(X = n) = (1 - p)^{n-1} p$. Note that the smallest value that n can be is 1, not 0. The first success can happen on the first attempt or later, but there has to be at least one attempt. **This formula is *not* given on the AP* Exam!**

- Returning to **Example 24:**

We want to find the probability that Julia makes her first free throw on her third attempt.

Applying the formula, we obtain:

$$P(x = 3) = (1 - .7)^{3-1} (.7) \approx 0.063$$

We can either use the formula to obtain the answer or we can use:

Geompdf(0.7,3) Notice that we drop the first value that we would have used in binompdf, which makes sense because in a geometric probability we don't have a fixed number of trials and that's what the first number in the binompdf command is used for.

Once again, show the work for the formula, not the calculator command. No credit will be given for calculator notation.

• **Example 25:** Using Example 24, what is the probability that Julia makes her first free throw on or before her fifth attempt?

Solution: This is again a geometric probability because Julia will keep shooting free throws until she makes one. For this problem, she could make the shot on her first attempt, second attempt, and so on until the fifth attempt. Applying the formula, we obtain:

$$P(X = 1) + P(X = 2) + P(X = 3) + P(X = 4) + P(X = 5)$$

or

$$(1 - .7)^0(.7)^1 + (1 - .7)^1(.7)^1 + (1 - .7)^2(.7)^1 + (1 - .7)^3(.7)^1 + (1 - .7)^4(.7)^1 \approx 0.9976$$

We could also use the following formula, which is the formula for finding the probability that it takes more than n trials to obtain the first success:

$$P(X > n) = (1 - p)^n$$

Using this formula and the concept of the complement, we obtain:

$$1 - P(X > 5) = 1 - (1 - .7)5 \approx 0.9976$$

Either method is OK as long as you show your work. If I used the first method, I would show at least three of the probabilities so that the grader of the AP* Exam knows that I understand how to apply the formula.

• **Example 26:** Using Example 24, find the expected value (mean) and the standard deviation.

Solution: The mean in this case is the expected number of trials that it would take before the first success is obtained. The formulas for the mean and standard deviation are:

$$\mu = \frac{1}{p} \quad and \quad \sigma = \sqrt{\frac{(1-p)}{p^2}}$$

Applying these formulas (which are not given on the AP formula sheet), we obtain:

$$\mu = \frac{1}{.7} \approx 1.4286$$

and

$$\sigma = \sqrt{\frac{(1-.7)}{.7^2}} \approx 0.7825.$$

Chapter

6

Sampling Distributions

6.1 Sampling Distributions

• Understanding sampling distributions is an integral part of inferential statistics. Recall that in **inferential statistics** you are making conclusions or assumptions about an entire population based on sample data. In this chapter, we will explore sampling distributions for means and proportions. In the remaining chapters, we will call upon the topics of this and previous chapters in order to study inferential statistics.

• From this point on, it's important that we understand the difference between a parameter and a statistic. A **parameter** is a number that describes some attribute of a **population**. For example, we might be interested in the mean, μ, and standard deviation, σ, of a population. There are many situations for which the mean and standard deviation of a population are unknown. In some cases, it is the population proportion that is not known. That is where inferential statistics comes in. We can use a statistic to estimate the parameter. A **statistic** is a number that describes an attribute of a **sample**. So, for the unknown μ we can use the sample mean, $\bar{x}$, as an estimate of μ. It's important to note that if we were to take another sample, we would probably get a different value for $\bar{x}$. In other words, if we keep sampling, we will probably keep getting different values for $\bar{x}$ (although some may be the same). Although μ may be unknown, it is a fixed number, as a population can have only one mean. The notation for the standard deviation of a sample is s. (Just remember that s is for "sample.") We sometimes use s to estimate σ, as we will see in later chapters.

- To summarize the notation, remember that the symbols μ and σ (parameters) are used to denote the mean and standard deviation of a population, and $\bar{x}$ and s (statistics) are used to denote the mean and standard deviation of a sample. You might find it helpful to remember that s stands for "statistic" and "sample" while p stands for "parameter" and "population." You should also remember that Greek letters are typically used for population parameters. Be sure to use the correct notation! It can help convince the reader (grader) of your AP* Exam that you understand the difference between a sample and a population.

- Consider again a population with an unknown mean, μ. Sometimes it is simply too difficult or costly to determine the true mean, μ. When this is the case, we then take a random sample from the population and find the mean of the sample, $\bar{x}$. As mentioned earlier, we could repeat the sampling process many, many times. Each time we would recalculate the mean, and each time we might get a different value. This is called **sampling variability**. Remember, μ does not change. The population mean for a given population is a fixed value. The sample mean, $\bar{x}$, on the other hand, changes depending on which individuals from the population are chosen. Sometimes the value of $\bar{x}$ will be greater than the true population mean, μ, and other times $\bar{x}$ will be smaller than μ. This means that $\bar{x}$ is an **unbiased estimator** of μ.

- The **sampling distribution** is the distribution of the values of the statistic if all possible samples of a given size are taken from the population. Don't confuse samples with sampling distributions. When we talk about sampling distributions, we are not talking about one sample; we are talking about **all possible samples** of a particular size that we could obtain from a given population.

• **Example 1:** Consider the experiment of rolling a pair of standard six-sided dice. There are 36 possible outcomes. If we define μ to be the average of the two dice, we can look at all 36 values in the sampling distribution of $\bar{x}$ (Figure 6.1). If we averaged all 36 possible values of $\bar{x}$, we would obtain the exact value of μ. This is always the case.

x-bar values

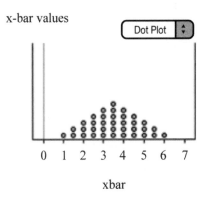

xbar

	1	2	3	4	5	6
1	(1,1)	(1,2)	(1,3)	(1,4)	(1,5)	(1,6)
2	(2,1)	(2,2)	(2,3)	(2,4)	(2,5)	(2,6)
3	(3,1)	(3,2)	(3,3)	(3,4)	(3,5)	(3,6)
4	(4,1)	(4,2)	(4,3)	(4,4)	(4,5)	(4,6)
5	(5,1)	(5,2)	(5,3)	(5,4)	(5,5)	(5,6)
6	(6,1)	(6,2)	(6,3)	(6,4)	(6,5)	(6,6)

Figure 6.1   Possible outcomes, $\bar{x}$, and dotplot when rolling a pair of dice.

• As mentioned, sometimes the value of $\bar{x}$ is below μ, and sometimes it is above μ. In Example 1, you can see that the center of the sampling distribution is exactly 3.5. In other words, the statistic, $\bar{x}$, is **unbiased** because the mean of the sampling distribution is equal to the true value of the parameter being tested, which is μ. Although the values of $\bar{x}$ may differ, they do not tend to consistently overestimate or underestimate the true mean of the population.

• As you will see later in this chapter, larger samples have less variability when it comes to sampling distributions. The spread is determined by how the sample is designed as well as the size of the sample. It's also important to note that the variability of the sampling distribution for a particular sample size does not depend on the size of the population from which the sample is obtained. An SRS (simple random sample) of size 4000 from the population of U.S. residents has approximately the same variability as an SRS of size 4000 from the population of Indiana residents. However, in order for the variability to be the same, both samples must be the same size and be obtained in the same manner. We want our samples to be obtained from correct sampling methods and the sample size to be large enough that our samples have low bias and low variability.

6.2 Sample Means and the Central Limit Theorem

• The following activity will help you understand the difference between a population and a sample, sampling distributions, sampling variability, and the Central Limit Theorem. I learned of this activity a few years ago from AP Statistics consultant and teacher Chris True. I am not sure where this activity originated, but it will help you understand the concepts presented in this chapter. If you've done this activity in class, that's great! Read through the next few pages anyway, as it will provide you with a good review of sampling and the Central Limit Theorem.

• The activity begins with students collecting pennies that are currently in circulation. Students bring in enough pennies over the period of a few days such that I get a total of about 600 to 700 pennies between all of my AP Statistics classes. Students enter the dates of the pennies into the graphing calculator (and Fathom) as they place the pennies into a container. These 600 to 700 pennies become our population of pennies. Then students make a guess as to what they think the distribution of our population of pennies will look like. Many are quick to think that the distribution of the population of pennies is approximately normal. After some thought and discussion about the dates of the pennies in the population, students begin to understand that the population distribution is not approximately normal but skewed to the left. Once we have discussed what we think the population distribution should look like, we examine a histogram or dotplot of the population of penny dates. As you can see in Figure 6.2, the distribution is indeed skewed to the left.

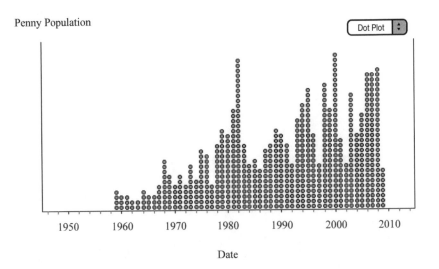

Figure 6.2 Population distribution for 651 pennies.

- It's important to discuss the shape, center, and spread of the distribution. As just stated, the shape of the distribution of the population of pennies is skewed left. The mean, which we will use as a measure of center, is $\mu = 1990.5868$. Since we are using the mean as the measure of center, it makes sense to use the standard deviation to measure spread. For this population of 651 pennies, $\sigma = 12.6937$ years.

- Once we've discussed the shape, center, and spread of the population distribution, we begin sampling. Students work in pairs and draw out several samples of each of the sizes: 4, 9, 16, 25, and 50. Sampling variability becomes apparent as students repeat samples for the various sample sizes. We divide up the sampling task among the students in class so that when we are done we have about 100 to 120 samples for each sample size. We graph the sampling distribution for each sample size and compute the mean and standard deviation of each sampling distribution.

- We can then analyze the sampling distribution for each sample size. We begin with $n = 4$ (Figure 6.3).

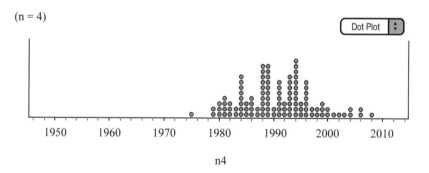

Figure 6.3 Sampling distribution for samples of size $n = 4$.

• Again, think about the shape, center, and spread of the distribution. Remember, this is a **sampling distribution.** This is the distribution of about 100 samples of size 4. As you can see in Figure 6.3, the shape of the sampling distribution is different from that of the population. Although the shape of the population is skewed left, the shape of the sampling distribution for $n = 4$ is more symmetrical. The center of the sampling distribution is $\mu_{\bar{x}} = 1990.6636$, which is very close to the population mean, μ. The spread of the sampling distribution is $\sigma_{\bar{x}} \approx 6.6031$. We can visualize that the spread of the sampling distribution is less than that of the population and that the mean of the sampling distribution (balancing point) is around 1990 to 1991. Note that if we had obtained all samples of size 4 from the population, then

$$\mu_{\bar{x}} = \mu = 1990.5868 \text{ and } \sigma_{\bar{x}} = \frac{\sigma}{\sqrt{n}} = \frac{12.6937}{\sqrt{4}} \approx 6.3469$$

Although it's impractical to obtain all possible samples of size 4 from the population of 651 pennies, our results are very close to what we would obtain if we had obtained all 7,414,857,450 samples. That's right; from a population of 651 pennies, the number of samples of size 4 you could obtain is

$$\binom{651}{4} = 7,414,857,450$$

• The following sampling distribution is for samples of size $n = 9$ (Figure 6.4).

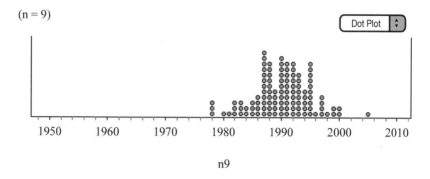

n9

Figure 6.4   Sampling distribution for samples of size $n = 9$.

• The sampling distribution for samples of size $n = 9$ is more symmetrical than the sampling distribution for $n = 4$. The mean and standard deviation for this sampling distribution are $\mu_{\bar{x}} \approx 1990.1339$ and $\sigma_{\bar{x}} \approx 4.8852$. Again, we can visualize that the mean is around 1990 to 1991 and that the spread is less for this distribution than that for $n = 4$.

• The following sampling distribution is for samples of size $n = 16$ (Figure 6.5).

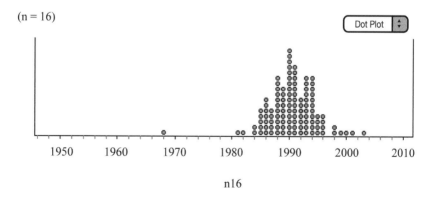

n16

Figure 6.5   Sampling distribution for samples of size $n = 16$.

• The sampling distribution for samples of size $n = 16$ is more symmetrical than the sampling distribution for $n = 9$. The mean and standard deviation for this sampling distribution are $\mu_{\bar{x}} \approx 1990.4912$ and $\sigma_{\bar{x}} \approx 4.3807$. Again, we can visualize that the mean is around 1990 to 1991 and that the spread is less for this distribution than that for $n = 9$.

• Notice the outlier of 1968. Although it's possible to obtain a sample of size $n = 16$ with a sample average of 1968 from our population of pennies, it is very unlikely. This is probably a mistake on the part of the student reporting the sample average or on the part of the student recording the sample average. It's interesting to note the impact that the outlier has on the variability of the sampling distribution. The theoretical standard deviation for the sampling distribution is

$$\sigma_{\bar{x}} = \frac{\sigma}{\sqrt{n}} = \frac{12.6937}{\sqrt{16}} \approx 3.1734$$

Notice that the standard deviation of our sampling distribution is greater than this value, which is due largely to the outlier of 1968. This provides us with a good reminder that the mean and standard deviation are not **resistant measures.** That is to say that they can be greatly influenced by extreme observations.

• The following is the sampling distribution obtained for $n = 25$ (Figure 6.6).

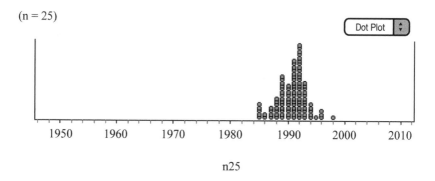

n25

Figure 6.6 Sampling distribution for samples of size $n = 25$.

- The mean and standard deviation for this sampling distribution are $\mu_{\bar{x}} \approx 1990.75$ and $\sigma_{\bar{x}} \approx 2.5103$. The shape of the sampling distribution is more symmetrical and more normal. We can visualize that the center of the distribution is again around 1990 to 1991 and that the variability is continuing to decrease as the sample size gets larger.

- The following is the sampling distribution obtained for $n = 50$ (Figure 6.7).

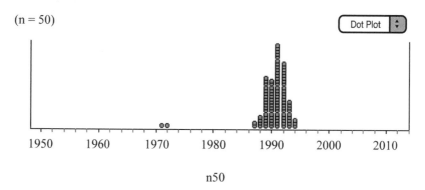

n50

Figure 6.7 Sampling distribution for samples of size $n = 50$.

- The mean and standard deviation for this sampling distribution are $\mu_{\bar{x}} \approx 1990.4434$ and $\sigma_{\bar{x}} \approx 3.0397$. The shape of the sampling distribution is more normal than that of any of the sampling distributions of smaller sample sizes. The center can again be visualized to be around 1990 to 1991, and the spread can be *visualized* to be *smaller* that that of the sampling distributions of smaller sample sizes. Notice, however, that the standard deviation of the sampling distribution is actually larger than that of size $n = 25$. How can this happen? Notice that there are two outliers. My students called them "super outliers." These are responsible for making the standard deviation of the sampling distribution larger than it would be theoretically. These outliers are very, very unlikely. We would be more likely to be struck by lightning twice while winning the lottery than to obtain two outliers as extreme as these. The outliers are probably due to human error in recording or calculating the sample means.

- The penny activity is the **Central Limit Theorem** (the Fundamental Theorem of Statistics) at work. The Central Limit Theorem says that as the sample size increases, the mean of the sampling distribution of $\bar{x}$ approaches a normal distribution with mean μ and standard deviation,

$$\sigma_{\bar{x}} = \frac{\sigma}{\sqrt{n}}$$

This is true for any population, not just normal populations! How large the sample must be depends on the shape of the population. The more non-normal the population, the larger the sample size needs to be in order for the sampling distribution to be approximately normal. Most textbooks consider 30 or 40 to be a "large" sample. The Central Limit Theorem allows us to use normal calculations when we are dealing with non-normal populations, provided that the sample size is large. It is important to remember that $\mu_{\bar{x}} \approx \mu$ and $\sigma_{\bar{x}} = \frac{\sigma}{\sqrt{n}}$ for any sampling distribution of the mean. The Central Limit Theorem states that the shape of the sampling distribution becomes more normal as the sample size increases.

6.3 Sample Proportions and the Central Limit Theorem

• Now that we've discussed sampling distributions, sample means, and the Central Limit Theorem, it's time to turn our attention to sample proportions. Before we begin our discussion, it's important to note that when referring to a sample proportion, we always use $\hat{p}$. When referring to a population proportion, we always use p. Note that some texts use π instead of p. In this case, π is just a Greek letter being used to denote the population proportion, not 3.1415 ...

• The Central Limit Theorem also applies to proportions as long as the following conditions apply:

1. The sampled values must be independent of one another. Sometimes this is referred to as the 10% condition. That is, the sample size must be only 10% of the population size or less. If the sample size is larger than 10% of the population, it is unlikely that the individuals in the sample would be independent.

2. The sample must be *large enough*. A general rule of thumb is that $np \geq 10$ and $n(1 - p) \geq 10$. As always, the sample must be random.

• If these two conditions are met, the sampling distribution of $\hat{p}$ should be approximately normal. The mean of the sampling distribution of $\hat{p}$ is exactly equal to p. The standard deviation of the sampling distribution is equal to:

$$\sqrt{\frac{p(1-p)}{n}}$$

• Note that because the average of all possible $\hat{p}$ values is equal to p, the sample proportion, $\hat{p}$, is an unbiased estimator of the population proportion, p.

- Also notice how the sample size affects the standard deviation.

Notice that as n gets larger, the fraction $\dfrac{p(1-p)}{n}$ gets smaller.

Thus, as the sample size increases, the variability in the sampling distribution decreases. This is the same concept discussed in the penny activity. Note also that for any sample size n, the standard deviation is largest from a population with $p = 0.50$.

Chapter

7

Inference for Means

7.1 The t-Distributions

• The **Central Limit Theorem (CLT)** is a very powerful tool, as was evident in the previous chapter. Our penny activity demonstrated that as long as we have a large enough sample, the sampling distribution of $\bar{x}$ is approximately normal. This is true no matter what the population distribution looks like. To use a z-statistic, however, we have to know the population standard deviation, σ. In the real world, σ is usually unknown. Remember, we use **statistical inference** to make predictions about what we believe to be true about a population.

• When σ is unknown, we estimate σ with s. Recall that s is the sample standard deviation. When using s to estimate σ, the standard deviation of the sampling distribution for means is $s_{\bar{x}} = \dfrac{s}{\sqrt{n}}$. When you use s to estimate σ, the standard deviation of the sampling distribution is called the **standard error** of the sample mean, $\bar{x}$.

• While working for Guinness Brewing in Dublin, Ireland, William S. Gosset discovered that when he used s to estimate σ, the shape of the sampling distribution changed depending on the sample size. This new distribution was not exactly normal. Gosset called this new distribution the *t-distribution*. It is sometimes referred to as the **student's t.**

• The t-distribution, like the standard normal distribution, is single-peaked, symmetrical, and bell shaped. It's important to notice, as mentioned earlier, that as the sample size (n) increases, the variability of the sampling distribution decreases. Thus, as the sample size increases, the t-distributions approach the standard normal model. When the sample size is small, there is more variability in the sampling distribution, and therefore there is more area (probability) under the density curve in the "tails" of the distribution. Since the area in the "tails" of the distribution is greater, the t-distributions are "flatter" than the standard normal curve. We refer to a t-distribution by its **degrees of freedom.** There are $n-1$ degrees of freedom. The "$n-1$" degrees of freedom are used since we are using s to estimate σ and s has $n-1$ degrees of freedom. Figure 7.1 shows two different t-distributions with 3 and 12 degrees of freedom, respectively, along with the standard normal curve. It's important to note that when dealing with a normal distribution,

$$z = \frac{\bar{x} - \mu}{\sigma / \sqrt{n}}$$ and when working with a t-distribution, $t = \frac{\bar{x} - \mu}{s / \sqrt{n}}$.

Using s to estimate σ introduces another source of variability into the statistic.

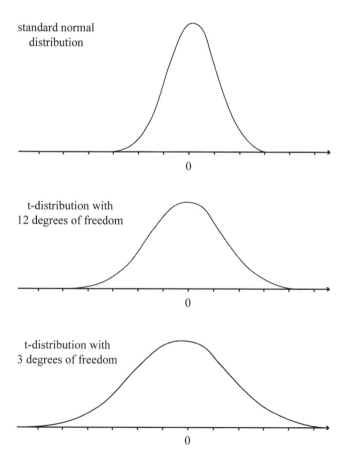

standard normal
distribution

0

t-distribution with
12 degrees of freedom

0

t-distribution with
3 degrees of freedom

0

Figure 7.1   Density curves with 3 and 12 degrees of freedom. Notice how the t-distribution approaches the standard normal curve as the degrees of freedom increases.

7.2 One-Sample t-Interval for the Mean

• As mentioned earlier, we use statistical inference when we wish to estimate some parameter of the population. Often, we want to estimate the mean of a population. Since we know that sample statistics usually vary, we will construct a **confidence interval.** The confidence interval will give a range of values that would be reasonable values for the parameter of interest, based on the statistic obtained from the sample.In this section, we will focus on creating a confidence interval for the mean of a population.

• When dealing with inference, we must always check certain **assumptions for inference.** This is imperative! These "assumptions" must be met for our inference to be reliable. We confirm or disconfirm these "assumptions" by checking the appropriate **conditions.** Throughout the remainder of this book, we will perform inference for different parameters of populations. We must always check that the assumptions are met before we draw conclusions about our population of interest. If the assumptions cannot be verified, our results may be inaccurate. For each type of inference, we will discuss the necessary assumptions and conditions.

• The assumptions and conditions for a one-sample t-interval or one-sample t-test are as follows:

Assumptions	Conditions
1. Individuals are independent	1. SRS and <10% of population ($10n$<N)
2. Normal population assumption	2. One of the following: – Given a normal population – Graph of sample data is symmetric with no outliers – Sample is large enough ($n \geq 30$) that the sampling distribution of $\bar{x}$ is approximately normal

- The t-procedures (t-interval and t-test) are **robust**, meaning that the results of our t-interval or t-test would not change very much even though the assumptions of the procedure are violated.

- Let's discuss the assumptions and conditions. The first assumption is that the individuals or observations are independent. This should be true if our sample data is an SRS or if our data comes from a randomized experiment and if the sample size is less than 10% of the population size. The second assumption is that the population is normal. We may know or be given that the population is normal. If this is the case, we state this in our problem. If we do not know or if we are not told that the population is normal and the sample size is small, we must then look at a graph of the sample data. A histogram or a modified boxplot is probably best suited for looking at the sample data. If the sample size is less than 30, we must be cautious of outliers or skewness in the data. Since normal distributions drop off quickly, it is unlikely to take a sample from a normal population and have the sample contain outliers or skewness. Outliers and strong skewness in a sample can be an indication that the population from which the sample is drawn might be non-normal. If the sample is large, we know that no matter what the population distribution looks like, we are guaranteed that the sampling distribution will be approximately normal. If you are asked to work on a problem for which the assumptions cannot be verified, state that this is the case and that the results of the inference being performed may be inaccurate.

- The following example involves finding a confidence interval. **As you solve statistical inference problems in this and the following chapters, keep in mind the following three steps:**

1. Identify the parameter of interest, choose the appropriate inference procedure, and verify that the assumptions and conditions for that procedure are met.
2. Carry out the inference procedure. Do the math! Be sure to apply the correct formula.
3. Interpret the results in the context of the problem.

• **Example 1:** Nolan wanted to estimate the average number of miles that a typical Indiana high-school male cross-country runner would run over a one-week period. The following is a random sample of the number of miles per week run by 20 male high-school cross-country runners in the state of Indiana. Find a 90% confidence interval for the average number of miles run per week for all male high-school cross-country runners in the state of Indiana.

20, 30, 35, 40, 40, 45, 45, 45, 50, 50, 50, 50, 52, 54, 55,
60, 60, 60, 70, 75

Solution:

Step 1: Find a 90% confidence interval for μ the mean number of miles run per week by a male high-school cross-country runner in the state of Indiana. Since σ is unknown, we will use a one-sample t-interval for the mean. We must check the assumptions and conditions.

Assumptions and conditions that verify:

1. **Individuals are independent.** We are given that the sample is random, and we can safely assume that there are more than 200 male cross-country runners in the state of Indiana ($10n < N$).

2. **Normal population assumption.** We are given a small sample, but a modified boxplot of the sample data appears to be symmetric with no outliers (Figure 7.2). We should be safe using t-procedures.

Weekly Mileage

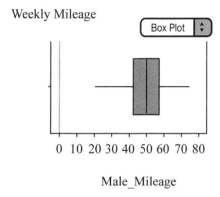

0 10 20 30 40 50 60 70 80

Male_Mileage

Figure 7.2 The sample data appears symmetrical with no outliers.

Step 2: Since the conditions are met, we will construct the confidence interval for the population mean, μ, using:

$$\overline{x} \pm t^*_{n-1} \times \frac{s}{\sqrt{n}} \text{ with } 20 - 1 = 19 \text{ } df$$

$t^* = 1.729$ can be found by using the t-distribution table. Use 19 df and cross-reference with 90% confidence at the bottom of the table. Study the table. Notice how the t^* values increase and approach z^* as the degrees of freedom and the sample size increase. Remember that the t-distributions approach the normal distribution when the sample size gets large.

$$49.3 \pm 1.729 \left(\frac{12.8968}{\sqrt{20}} \right)$$

$$\left(44.314, \ 54.286 \right)$$

Step 3: Conclude in context! We are 90% confident that the true average number of miles that Indiana high-school male cross-country runners run in a given week is between 44.314 and 54.286 miles ("true average" refers to the average of *all* Indiana male cross-country runners).

• Notice that the sample mean, $\bar{x}$, is the center of the confidence interval. The distance between the ends of the confidence interval and the center is called the margin of error. Thus, $t^*_{n-1} \times \dfrac{s}{\sqrt{n}}$ is the margin of error.

• I highly recommend using the three-step process when doing inference. Some textbooks write up the inference problems a little differently. As long as you have all the essentials of the inference procedure, it doesn't make a huge difference how you organize it. I have found it best to find a system you can use so that you don't leave out any of the essentials of inference, including the assumptions and conditions. Always show your work!

• Refer to Example 1; it's important to note several things. First of all, make sure you define any variables you use. State what procedure you are going to use and, of course, make sure you check the assumptions and conditions. If you refer to a histogram, modified boxplot, or any type of graph of the data, make sure to include the graph. Don't assume that the reader (grader) will know what you are talking about. Always plot the data if you are given a small sample. If the sample is over 30 (some books say 40), then you do not have to plot the sample data. Remember that if the sample is "large enough," then the sampling distribution should be approximately normal, no matter what the sample data looks like. If the population is given to be normal, you don't have to worry about skewness or outliers either. If you are not given a normal population and the sample is less than 30, then you must plot the sample data to look for skewness and outliers.

• Note that a graphing calculator could be used to find the confidence interval. It can also perform various tests of significance. The graphing calculator is a powerful tool, but it doesn't take the place of applying formulas and showing our work.

Interpreting Confidence Intervals

• It is highly likely that your understanding of how to interpret a confidence interval will be tested on the AP* Exam. What exactly can we say when we interpret the confidence interval in the context of the problem? In Example 1, we concluded, with 90% confidence, that μ was between 44.314 and 54.286 miles. That is, the average number of miles run by a typical male high-school cross-country runner in the state of Indiana is between 44.314 and 54.286. What exactly does this mean? Here's what we can say: *We can say* that if this process were repeated many times, approximately 90% of all confidence intervals that we construct would contain the true mean. That is, if we were to obtain 100 different samples, find the mean of each sample, and construct 100 different confidence intervals, we would expect about 90 of them to contain the true population mean, μ. That is also to say that about 10 of our confidence intervals would not contain the true population mean. No matter how carefully we obtain our random sample, there will always be sampling variability, and this variability makes the process imperfect. *Be Careful! We cannot say* that there is a 90% probability that the true mean is between 44.314 and 54.286 miles. *We cannot say* that 90% of all males cross-country runners in the state of Indiana run between 44.314 and 54.286 miles per week on average. These and comments like these are common on multiple-choice questions on exams. *We can only say* that if this process were repeated many times, 90% of all confidence intervals that we construct would contain the true population mean (Figure 7.3).

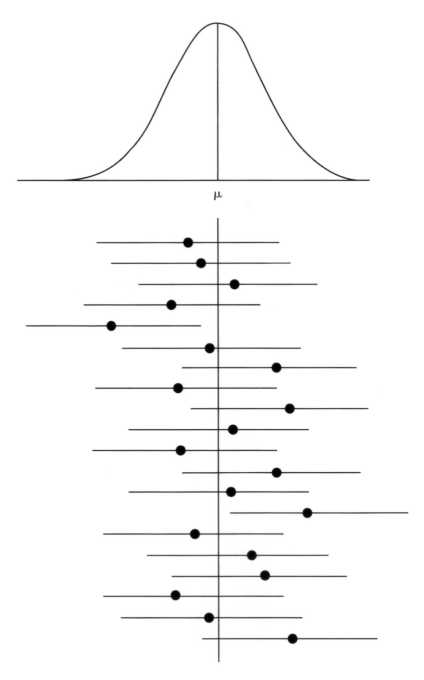

Figure 7.3   18 out of 20 confidence intervals contain the true
population mean.

7.3 One-Sample t-Test for the Mean

• The following example (Example 2) will be used to outline the essentials of a one-sample t-test. This is an example of a **hypothesis test**, or **test of significance.** We use this form of statistical inference when we wish to test a claim that has been made concerning a population. As with confidence intervals, we use sample data to help us make decisions about the population of interest. In other words, we use the sample data to see if there is enough "evidence" to support the claim or to reject it.

• We will use the same basic three-step method for hypothesis testing that we used for confidence intervals with some minor modifications. Remember, it's not the numbering of the steps that's important; it's what's in the three steps. Make sure that no matter how you solve inference problems, you include all the essentials.

• Use the following three-step method when performing a hypothesis test for the mean of a population:

1. Identify the parameter of interest, choose the appropriate inference procedure, and verify that the assumptions and conditions for that procedure are met. Define any variables of interest. State the appropriate null and alternative hypotheses.

2. Carry out the inference procedure. Do the math! Calculate the test statistic and find the p-value.

3. Interpret the results in context of the problem. This is by far the most important part of inference. Be sure that your decision to reject or fail to reject the null hypothesis is done in context of the problem and is based upon the p-value.

• As noted in step 1, hypothesis testing typically involves a **null hypothesis** and an **alternative hypothesis.** It's important to note that we are not proving anything; we are simply testing to see if there is enough evidence to reject or fail to reject the null hypothesis. The null hypothesis is denoted by H_0, pronounced *H-nought*. The alternative hypothesis is denoted by H_a.

• The null hypothesis should always include an equality (*like* $\leq$, =, *or* $\geq$) and must always be written using parameters and not statistics! Of course, you should define any variables you use. For example: $H_0 : \mu = \mu_0$, where μ_0 is the hypothesized value.

• The alternative hypothesis can be one-sided or two-sided. A one-sided alternative would be either $H_a : \mu < \mu_0$ *or* $H_a : \mu > \mu_0$. A two-sided alternative would be: $H_a : \mu \neq \mu_0$.

• The same assumptions and conditions must be met for a hypothesis test as for a confidence interval. Remember: Always check the assumptions and conditions!

• When the conditions for using one-sample t-procedures are met, we can use the test statistic $t = \dfrac{\bar{x} - \mu}{s/\sqrt{n}}$, with $n-1$ degrees of freedom.

• We use the **p-value** to determine if we reject or fail to reject the null hypothesis. The p-value is the probability of obtaining a sample statistic as extreme or more extreme than we have obtained, *given that the null hypothesis is true*. The smaller the p-value, the more evidence we have to reject the null hypothesis. Most graphing calculators will calculate the p-value for us, but it's important that we understand how it is calculated. We will discuss how the p-value is calculated after we've completed Example 2.

• **Example 2:** A recent news broadcast stated that the typical U.S. teen plays an average of 15 hours of video games per week. A group of parents, at a meeting discussing the broadcast, believe that teens actually play more than 15 hours of video games per week. A random sample of 42 teens is selected, and it is determined that the average number of hours that video games are played is 16.5 hours with a standard deviation of 4.5 hours. Is there evidence to support the parents' claim at the 5% level of significance?

Solution:

Step 1: We will conduct a one-sample t-test.

> Let μ = *mean number of hours that a typical teen plays video games per week.*

$H_0 : \mu = 15$

$H_a : \mu > 15$

Assumptions and conditions that verify:

1. **Individuals are independent.** We are given that the sample is random, and we can safely assume that there are more than 420 teenagers in the U.S. ($10n < N$).

2. **Normal population assumption.** We are given a large sample; therefore the sampling distribution of $\bar{x}$ should be approximately normal. We should be safe using t-procedures.

Step 2: $t = \dfrac{\bar{x} - \mu}{s/\sqrt{n}} = \dfrac{16.5 - 15}{4.5/\sqrt{42}} \approx 2.1602 \quad df = 41$

$p \approx 0.0183$

Step 3: With a p-value of 0.0183, we reject the null hypothesis at the 5% level. There appears to be enough evidence to reject the null hypothesis and conclude that the typical teen plays more than 15 hours of video games per week.

• Let's return to the p-value. What does a p-value of 0.0183 really mean? Think about it this way. If the typical teen really does play an average of 15 hours of video games a week, the probability of taking a random sample from that population and obtaining an $\bar{x}$ value of 16.5 or more is only 0.0183. In other words, it's possible, but pretty unlikely. Only about 1.83% of the time can we obtain a sample average of 16.5 hours or greater, *if* the true population mean is 15 hours.

• In Example 2, we rejected the null hypothesis at the 5% level (this is called the alpha level, α). The most common levels at which we reject the null hypothesis are the 5% and 1% levels. That's not to say that we can't reject a null hypothesis at the 10% level or even at the 6% or 7% levels; it's just that 1% and 5% happen to be commonly accepted levels at which we reject or fail to reject the null hypothesis.

• You may struggle a little while first using the p-value to determine whether you should reject or fail to reject the null hypothesis. Always compare your p-value to the given $\alpha - level$. In Example 2, we used an $\alpha - level$ of 0.05. Our p-value of 0.0183 led us to reject at the 5% level because 0.0183 is less than 0.05. We did not reject at the 1% level because 0.0183 is greater than 0.01. To reject at a given $\alpha - level$, the p-value must be less than the $\alpha - level$.

• If an $\alpha - level$ is not given, you should use your own judgment. You are probably safe using a 1% or 5% alpha level. However, don't feel obligated to use a level. You can make a decision based on the p-value without using an alpha level. Just remember that the smaller the p-value, the more evidence you have to reject the null hypothesis.

• The p-value in Example 2 is found by calculating the area to the right of the test statistic $t = 2.1602$ under the t-distribution with $df = 41$. If we had used a two-sided alternative instead of a one-sided alternative, we would have obtained a p-value of 0.0367, which would be double that of the one-sided alternative. Thus, the p-value for the two-sided test would be found by calculating the area to the right of $t = 2.1602$ and combining that with the area to the left of $t = -2.1602$.

7.4 Two-Sample t-Interval for the Difference Between Two Means

• We are sometimes interested in the difference in two population means, $\mu_1 - \mu_2$. The assumptions and conditions necessary to carry out a confidence interval or test of significance are the same for two-sample means as they are for one-sample means, with the addition that the samples must be independent of one another. You must check the assumptions and conditions for each independent sample.

• The assumptions and conditions for a two-sample t-interval or two-sample t-test are as follows:

Assumptions	Conditions
1. Samples are independent of each other	1. Are they? Does this seem reasonable?
2. Individuals in each sample are independent	2. Both SRSs and both <10% population ($10n$<N for both samples)
3. Normal populations assumption	3. One of the following: – Given normal populations – Graph of data for both samples shows no outliers or strong skewness – Samples are both large ($n \geq 30$); therefore the sampling distribution of $\bar{x}_1 - \bar{x}_2$ is approximately normal

• Remember that the mean of the sampling distribution of $\bar{x}_1 - \bar{x}_2$ is $\mu_1 - \mu_2$.

• The standard deviation of the sampling distribution is $\sqrt{\dfrac{\sigma_1^2}{n_1} + \dfrac{\sigma_2^2}{n_2}}$.

Remember that the population standard deviations are usually unknown. Recall that when this is the case, we use the sample standard deviation to estimate the population standard deviation. Thus, the **standard error**

(SE) of the sampling distribution is $\sqrt{\dfrac{s_1^2}{n_1} + \dfrac{s_2^2}{n_2}}$.

• Once we've checked the assumptions and conditions, we can proceed to finding the confidence interval for the difference of the means of

the two independent groups. We can use $\left(\bar{x}_1 - \bar{x}_2 \right) \pm t^* \times \sqrt{\dfrac{s_1^2}{n_1} + \dfrac{s_2^2}{n_2}}$.

The t^* value depends on the particular level of confidence that you want and on the degrees of freedom (df).

• To find the degrees of freedom of a two-sample t-statistic, we can use one of two methods:

Method 1: Use the calculator-generated degrees of freedom. This gives an accurate approximation of the t-distribution based on degrees of freedom from the data. Usually, we obtain non-whole number values using this method. The formula our calculator uses is somewhat complex, and we probably don't need to be too concerned with how the degrees of freedom are calculated. Make sure, however, that you always state the degrees of freedom that you are using, regardless of what method you use.

Method 2: Use the degrees of freedom equal to the smaller of the two values of $n_1 - 1$ and $n_2 - 2$. This is considered a conservative method.

• **Example 3:** Two high-school cross-country coaches from different teams are discussing their boys' and girls' teams. One coach believes that male and female cross-country runners in the state of Indiana differ in the number of miles they run, on average, each week. The other coach disagrees. He feels that male and female cross-country athletes run about the same number of miles per week, on average. Construct a 95% confidence interval for the difference in average weekly mileage between male and female cross-country runners in the state of Indiana. Consider the data obtained from two independent random samples:

Boys: 20, 30, 35, 40, 40, 45, 45, 45, 50, 50, 50, 50, 52, 54, 55, 60, 60, 60, 70, 75

Girls: 20, 20, 30, 35, 35, 40, 40, 40, 45, 45, 45, 50, 50, 50, 52, 60, 60, 60, 60, 60

Solution:

Step 1:

Let μ_1 = *mean weekly mileage for male runners.*

Let μ_2 = *mean weekly mileage for female runners.*

Find the mean difference, $\mu_1 - \mu_2$, in weekly mileage between male and female high-school cross-country runners in the state of Indiana.

Assumptions and conditions that verify:

Samples are independent of one another. We are given that the samples are independent of one another.

Individuals are independent. We are given that the samples are random, and we can safely assume that there are more than 200 male and 200 female cross-country runners in the state of Indiana ($10n <$ N for both samples).

Normal population assumption. We are given small samples, but modified boxplots for each sample appear to be symmetric with no outliers. We should be safe using t-procedures.

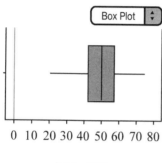

Male_Mileage

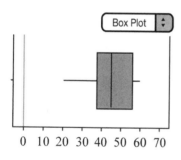

Female_Mileage

Figure 7.4 Modified boxplots show neither outliers nor strong skewness.

Step 2: Since the assumptions and conditions have been met, we will construct a two-sample t-interval.

$$\left(\overline{x}_1 - \overline{x}_2\right) \pm t^* \times \sqrt{\frac{s_1^2}{n_1} + \frac{s_2^2}{n_2}}$$

$$\left(49.3 - 44.85\right) \pm 2.042 \times \sqrt{\frac{12.8968}{20} + \frac{12.5626}{20}}$$

$df = 37.9738$ from calculator

$(-3.7, 12.6)$

Note that it would also be acceptable to use 19 degrees of freedom, as it is the smaller $n-1$.

Step 3: We are 95% confident that the true difference in the means of male and female high-school cross-country runners in the state of Indiana is between −3.7 miles and 12.6 miles.

• Does our confidence interval shed any light on the coaches' discussion concerning whether male and female runners differ in weekly average mileage? We can see from our interval that males run between −3.7 miles to 12.6 miles more than their female counterparts. Think about it! What does it mean for male runners to run −3.7 miles more than females? It means that they are running 3.7 miles *less* than females. Equally important is the fact that *zero* is contained in the confidence interval, which means that male and female cross-country runners in the state of Indiana might run the same number of weekly miles, on average. Remember, we are not 100% sure of anything. Recall that a 95% confidence interval implies that if we were to construct many, many confidence intervals using the same process, about 95 out of every 100 confidence intervals would contain the true mean difference in the amount of miles run by male and female high-school cross-country runners in the state of Indiana.

• In the next section, we will perform a two-sample t-test using the same data as the previous problem. We will outline the appropriate steps of inference and then show how the confidence interval relates to the significance test.

7.5 Two-Sample t-Test for the Difference Between Two Means

• The assumptions and conditions for a two-sample hypothesis test for means are the same as the assumptions and conditions for a two-sample t-interval. The null hypothesis for this type of test can be written as:

$H_0 : \mu_1 = \mu_2 \quad or \quad H_0 : \mu_1 - \mu_2 = 0$

As with a one-sample t-test, the alternative hypothesis can be written with $\neq$, $<$, or $>$. Once the appropriate assumptions and conditions have been met, we can calculate the two-sample t-statistic as follows:

$$t = \frac{\left(\overline{x_1} - \overline{x_2}\right) - \left(\mu_1 - \mu_2\right)}{\sqrt{\dfrac{s_1^2}{n_1} + \dfrac{s_2^2}{n_2}}}$$

• **Example 4:** Let's revisit Example 3. Two cross-country coaches from different teams are discussing their boys' and girls' teams. One coach believes that male and female cross-country runners in the state of Indiana differ in the number of miles they run, on average, each week. The other coach disagrees. He feels that male and female cross-country athletes run about the same number of miles per week, on average. Is there reason to believe that male and female cross-country runners in Indiana differ in the number of miles they run, on average, each week? Give appropriate statistical evidence to support your answer.

Solution:

Step 1: To answer the question, we will perform a two-sample t-test. We have already defined our variables and checked the appropriate assumptions and conditions for this type of inference in Example 3. We state the null and alternative hypotheses:

$H_0 : \mu_1 = \mu_2$

$H_0 : \mu_1 \neq \mu_2$

Step 2: Since the assumptions and conditions have been met, we can calculate the test statistic as follows:

$$t = \frac{\left(\overline{x}_1 - \overline{x}_2\right) - \left(\mu_1 - \mu_2\right)}{\sqrt{\dfrac{s_1^2}{n_1} + \dfrac{s_2^2}{n_2}}}$$

$$t = \frac{\left(49.3 - 44.85\right) - 0}{\sqrt{\dfrac{12.8968}{20} + \dfrac{12.5626}{20}}} \quad df = 37.9738$$

$t \approx 1.1053 \quad p \approx .2760$ **(p-value)**

Step 3: With a p-value of approximately 0.2760, we fail to reject the null hypothesis at any reasonable level of significance. We conclude that male and female cross-country runners do not differ in average weekly mileage.

• Consider what the p-value really means in this case. If the null hypothesis were really true—that is, male and female cross-country runners in Indiana really do not differ in weekly mileage—the probability of obtaining two samples with average values as different as those obtained is approximately 27.60%. In other words, more than 25% of the time, when sampling, we would obtain sample values from these populations that are as different as or more different than we have obtained. There is not enough statistical evidence to support the claim that males and females differ in weekly average mileage. Remember, typical p-values that lead to rejection of the null hypothesis are usually 10% or less.

• In the previous section (Example 3), we constructed a 95% confidence interval for the difference in mean weekly mileage between the male and female cross-country runners. Recall that the 95% confidence interval we obtained in Example 3 contained zero. If the difference between μ_1 and μ_2 were really zero, we could conclude that there was no difference between the means. That is exactly what our results from the significance test in Example 4 are telling us.

• Understand the connection between a confidence interval and a test of significance. A 95% confidence interval is equivalent to a two-sided significance test at the 5% level. A 90% confidence interval is equivalent to a two-sided significance test at the 10% level or to a one-sided test at the 5% level. Make sure you understand the connection between confidence intervals and tests of significance.

7.6 Matched Pairs (One-Sample t)

• Recall Example 4 from Chapter 4, concerning a matched-pairs experiment. A manufacturer of bicycle tires wants to test the durability of a new material used in bicycle tires. A completely randomized design might be used where one group of cyclists uses tires made with the "old" material, and another group uses tires made with the "new" material. The manufacturer realizes that not all cyclists will ride their bikes on the same type of terrain and in the same conditions. To help control these variables, we can implement a **matched-pairs design.** Recall that matching is a form of blocking. One way to do this is to have each cyclist use both types of tires. A coin could be used to determine whether the cyclist uses the tire with the new material on the front of the bike or on the rear. We could then compare the front and rear tires for each cyclist.

• Suppose in this experiment that the researcher has kept track of the number of miles that each tire lasted before needing to be replaced. We might consider looking at the average number of miles that the tires with the new material lasted and compare that with the average number of miles that the tires with the old material lasted. However, we should realize that the assumption of independence has been violated because each cyclist is using one tire with the old material and one tire with the new material. Since the data comes from matching, the data sets are not independent. When this happens, we take the difference for each pair of data and use a *one-sample t-procedure*, instead of a two-sample t-procedure. **Remember: Matched pairs are always a one-sample t-procedure, not a two-sample t-procedure!**

• **Example 5:** Find a 90% confidence interval for the mean difference in the mileage obtained for tires with the new material and tires with the old material. Following are the paired differences (new minus old) for each of the 17 riders, chosen at random, who took part in the experiment:

50, 45, 50, 50, 100, 100, 100, −10, 75, −25, 0, 25, 75, 25, 50, 40, 35

Solution:

Step 1: We want to estimate μ, the mean difference in the number of miles obtained from tires using the new material and tires using the old material. It's common to use μ_d or μ_{diff} to show that we are interested in the mean difference. Since we are using data from a matched-pairs experiment, we will check the assumptions and conditions for a one sample t-interval.

Assumptions and conditions that verify:

Individuals are independent. We are given that the sample is random. We can safely assume that there are more than 170 cyclists who might use these types of tires ($10n < N$).

Normal population assumption. We are given a small sample, but a modified boxplot of the *differences* (Figure 7.5) appears to be fairly symmetric with no outliers. We should be safe using t-procedures.

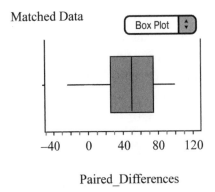

Matched Data   Box Plot ⬍

$$-40 \quad 0 \quad 40 \quad 80 \quad 120$$

Paired_Differences

Figure 7.5   A modified boxplot of the differences
shows little skewness and no outliers.

Step 2: Since the assumptions and conditions are met, we will construct
a one-sample t-interval for the differences.

$$\overline{x} \pm t^{*}_{n-1} \times \frac{s}{\sqrt{n}}$$

$$46.1765 \pm 1.746 \times \frac{36.9345}{\sqrt{17}} \quad df = 16$$

$(30.537, 61.816)$

Step 3: We are 90% confident that the mean difference in the number
of miles that tires manufactured with the new material lasted compared
to those constructed with the old material is between 30.537 and
61.816 miles.

7.7 Errors in Hypothesis Testing: Type I, Type II, and Power

• No matter how carefully we set up an experiment or how carefully we obtain a sample, we can always make a mistake when conducting a hypothesis test. Due to sampling variability, we sometimes reject the null hypothesis when we should fail to reject it and sometimes fail to reject the null hypothesis when we should reject it. These types of mistakes are called type I and type II errors.

A **type I error** occurs when we reject the null hypothesis when, in fact, it is actually correct. The probability of making a type I error is equal to the significance level ($\alpha - level$) of the test.

A **type II error** occurs when we fail to reject the null hypothesis when, in fact, the null hypothesis is false. The probability of a type II error is referred to as β.

• Finding the probability of a type I error is as simple as stating the significance level of the test. Students are not responsible for calculating the probability of a type II error, but you should understand the concepts of both type I and type II errors and be able to explain them in context of the problem. You should also understand that the probability of a type II error is dependent upon the chosen alternative value of the given parameter.

• The **power** of the test is the probability of rejecting the null hypothesis given that a particular alternative value is true. The power of the test is equal to $1 - \beta$. We want the power of the test to be relatively high. Think of it this way: If the null hypothesis is not true and a particular alternative value is true, then we *should be* rejecting the null hypothesis. Typically, we want the power of the test to be around 80% or above.

• **There are four ways we can increase the power of a hypothesis test.**
The first two are probably the most important for you to remember.

1. **Increase the α – level.** Increasing the α – *level* is one method of increasing the power of the test.

2. **Increase the sample size.** Increasing the sample size makes us more confident about our decision to reject or fail to reject the null hypothesis. Cost sometimes prohibits increasing the sample size.

3. **Decrease the standard deviation.** Sometimes it is possible to decrease the standard deviation, and sometimes it is not. Machinery can sometimes be finely tuned so that the production of goods can be more precise, which can in turn reduce the variability.

4. **Choose a different alternative value.** Choosing an alternative value that is further away from the value of the null hypothesis will increase the value of the power. The power is affected by the difference between the hypothesized value and the true value of the parameter.

Chapter

8

Inference for Proportions

8.1 One-Sample z-Interval for Proportions

• Now that we've discussed various statistical inference procedures for population means, it's time to turn our attention to statistical inference involving proportions. We are often concerned about the unknown proportion of the population that has some particular outcome of interest.

• Remember the appropriate statistical notation when dealing with proportions. Always use $\hat{p}$ when referring to a sample proportion and p when referring to a population proportion.

• As discussed in Chapter 6, the sampling distribution of $\hat{p}$ is approximately normal, provided that np and $n(1-p)$ are at least 10. The standard deviation of the sampling distribution of $\hat{p}$ is

$\sqrt{\dfrac{p(1-p)}{n}}$ as long as the population is at least 10 times the sample size.

When dealing with confidence intervals, we do not know p. Because $\hat{p}$ is an unbiased estimator of p, we use $\hat{p}$ to estimate p. These two values should be close in value, provided that the sample is large enough.

We can then use the standard error of $\hat{p}$, which is: $SE = \sqrt{\dfrac{\hat{p}(1-\hat{p})}{n}}$.

• When constructing a one-proportion z-interval, we use:

$$\hat{p} \pm z^* \sqrt{\frac{\hat{p}(1-\hat{p})}{n}}$$

• As with any type of inference, always check the assumptions and conditions of the inference procedure. The assumptions and conditions for a one-proportion z-interval or test are:

Assumptions	Conditions
1. Individuals are independent	1. SRS and $n < 10\%$ of population
2. Sample is large enough	2. $np \geq 10$ and $n(1-p) \geq 10$ Use $\hat{p}$ for C.I. and p_0 for tests

• To ensure that we include all the essentials of inference, we will follow the 3-step method used in Chapter 7.

1. Identify the parameter of interest, choose the appropriate inference procedure, and verify that the assumptions and conditions for that procedure are met.

2. Carry out the inference procedure. Do the math! Be sure to apply the correct formula.

3. Interpret the results in context of the problem.

• **Example 1:** Cassidy is interested in knowing the percentage of fourth graders in the Indianapolis area who own a cell phone in hopes of convincing her parents that she should own one too. With the help of her favorite statistician, she gathers information from a random sample of 100 fourth graders in the Indianapolis area. She finds that 18 of the 100 fourth graders sampled do indeed own a cell phone. Construct a 90% confidence interval for the true proportion of fourth graders who own cell phones.

Solution:

Step 1: We want to estimate p, the true proportion of fourth graders who own cell phones. We must check the assumptions and conditions.

Assumptions and conditions that verify:

1. **Individuals are independent.** We are given a random sample, and we are safe to assume that there are more than 1000 fourth graders in the Indianapolis area ($10n<N$).

2. **Sample is large enough:** $\hat{p} = \dfrac{18}{100} = 0.18$

   $100(.18) = 18 \geq 10$ and $100(.82) = 82 \geq 10$. Hence, we are safe to assume that the sampling distribution of $\hat{p}$ is approximately normal.

Step 2: Since the assumptions and conditions for inference have been met, we will construct the one-proportion z-interval.

$$\hat{p} \pm z^* \sqrt{\frac{\hat{p}(1-\hat{p})}{n}}$$

$$0.18 \pm 1.645 \sqrt{\frac{.18(1-.18)}{100}}$$

$$\left(0.11681, 0.24319\right)$$

Step 3: We are 90% confident that the true proportion of fourth graders who own cell phones in the Indianapolis area is between 11.68% and 24.32%. Cassidy will remain one of those "unfortunate" fourth graders who do not own a cell phone.

Margin of Error

• Now that we've discussed how to construct a one-sample t-interval for the mean of a population and a one-proportion z-interval for the population proportion, it's time to discuss the **margin of error**. When dealing with a one-proportion z-interval, the margin of error is the distance from the endpoints of the confidence interval to the center of the interval, $\hat{p}$. The margin of error is the product of the z* value and the standard error and is affected primarily by the sample size and the z* value (confidence level). The margin of error for a t-interval is affected in a similar fashion by the sample size and the level of confidence.

• We know that as the sample size increases, the variability of the sampling distribution decreases. The effects of changing the sample size on the confidence interval become evident if we change the sample size while keeping the standard deviation and confidence level the same. Consider Example 1: How does the confidence interval change when we increase the sample size in Example 1 from 100 to 500? What happens if we increase the sample size in Example 1 to 1000?

90% C.I. $n = 100$ $0.18 \pm 1.645 \sqrt{\dfrac{.18(1-.18)}{100}}$ $(0.1168, 0.2432)$

90% C.I. $n = 500$ $0.18 \pm 1.645 \sqrt{\dfrac{.18(1-.18)}{500}}$ $(0.1517, 0.2083)$

90% C.I. $n = 1000$ $0.18 \pm 1.645 \sqrt{\dfrac{.18(1-.18)}{1000}}$ $(0.1600, 0.2000)$

Notice that as the sample size increases, the width of the confidence interval decreases. This is due to the fact that there is less sampling variability in larger samples than in smaller samples. Thus, the standard deviation of the sampling distribution is smaller, and, consequently, the margin of error is smaller. This causes the confidence interval to be narrower. It's easy to see the advantage of using larger samples when performing inference. Cost and other factors sometimes prohibit using larger samples.

• What about the level of confidence for the interval? How does the confidence interval in Example 1 change if we keep the sample size of 100 but change the level of confidence to 95%? What happens to the confidence interval if we change the level of confidence to 99%?

90% C.I. $n = 100$ $0.18 \pm 1.645 \sqrt{\dfrac{.18(1-.18)}{100}}$ $(0.1168, 0.2432)$

95% C.I. $n = 100$ $0.18 \pm 1.960 \sqrt{\dfrac{.18(1-.18)}{100}}$ $(0.1047, 0.2553)$

99% C.I. $n = 100$ $0.18 \pm 2.576 \sqrt{\dfrac{.18(1-.18)}{100}}$ $(0.0810, 0.2790)$

Notice that as the confidence level increases, so does the width of the confidence interval. Mathematically, this is the result of using a different value for z^*. Recall that z^* is the number of standard deviations from the mean. To be more confident that our interval contains the true population proportion, the interval must be wider. Realize that the cost of increasing the level of confidence is that the interval becomes wider as the level increases.

• We are sometimes required to find the sample size needed to obtain a desired margin of error. The next example illustrates how to find the needed sample size when dealing with proportions. A similar method is used to find the sample size when dealing with means.

• **Example 2:** A polling organization wants to determine the sample size needed to estimate p, the proportion of voters who plan to vote for a particular candidate. The organization wants to estimate p with 98% confidence and a margin of error of no more than 3%. How large of a sample is needed?

Solution: We can use the formula for the margin of error.

$$z^* \sqrt{\frac{p^*(1-p^*)}{n}} \leq m$$

Using 2.326 (98% confidence) for z and 0.5 for p^*

$$2.326 \sqrt{\frac{(.5)\ (1-.5)}{n}} \leq 0.03$$

Solving for n, we obtain:

$$n \geq 1502.8544$$

Because the number of people surveyed must be a whole number, we round to 1503. It should be noted that 0.5 is used for p^*. If previous sampling had been done and an estimate of p had been obtained, that estimate could be used instead of 0.5. However, 0.5 will always give a sample size that is larger than any other value used for p^*. So if in doubt, use 0.5.

• As mentioned earlier, we can find a desired sample size needed for a particular margin of error when dealing with means in the same manner as we do proportions. Just use the part of the formula that comes after the $\pm$ in the appropriate confidence interval and solve for n. Always round your answer up to the next whole number.

8.2 One-Sample z-Test for Proportions

• Hypothesis testing for a one-proportion z-test is similar to that of a one-sample t-test, at least to some extent. The difference is that we are dealing with proportions instead of means. The assumptions and conditions are the same for a one-proportion z-test as they are for a one-proportion z-interval. Keep in mind, however, that since we do not know the true population proportion, p, we use the hypothesized value, p_0, when checking the assumptions and conditions. We also use p_0 for calculating the standard error of the sampling distribution of $\hat{p}$. As with a one-sample t-test, we use an equality when stating the null hypothesis and an inequality when stating the alternative hypothesis. We will use the same three-step process for organizing the inference procedure as we have done thus far to help ensure that we include the essentials of inference.

• Provided that the assumptions and conditions for a one-proportion z-test are met, we can calculate the test statistic using:

$$z = \frac{(\hat{p} - p_0)}{\sqrt{\dfrac{p_0(1 - p_0)}{n}}}$$

We then obtain a p-value based on the value of z and make a decision whether to reject or fail to reject the null hypothesis. Consider the following example.

- **Example 3:** A beverage company claims that 45% of adults drink diet soda. Skeptical about the claim, Addison obtains a random sample of 1000 adults and finds that 419 of them drink diet soda. Is there evidence to support Addison's suspicion that less than 45% of adults drink diet soda?

Solution:

Step 1: We will conduct a one-proportion z-test.

Let p = *proportion of all adults who drink diet soda*

$H_0 : p = 0.45$

$H_a : p < 0.45$

Assumptions and conditions that verify:

1. **Individuals are independent.** We are given a random sample, and we are safe to assume that there are more than 10,000 adults who drink diet soda ($10n < N$).

2. **Sample is large enough:** $\hat{p} = \dfrac{419}{1000} = 0.419$

   $1000(.419) = 419 \geq 10$ and $1000(.581) = 581 \geq 10$. Be sure to show the actual numbers! Therefore, we are safe to assume that the sampling distribution of $\hat{p}$ is approximately normal.

Step 2: The assumptions and conditions for inference have been met; we will perform a one-proportion z-test.

$$z = \frac{(\hat{p} - p_0)}{\sqrt{\dfrac{p_0(1 - p_0)}{n}}}$$

$$z = \frac{(.419 - .45)}{\sqrt{\dfrac{.419(1 - .419)}{1000}}}$$

$$z \approx -1.9705 \quad p \approx 0.0244$$

Step 3: With a p-value of 0.0244, we reject the null hypothesis at the 5% level. We conclude that the proportion of adults who drink diet soda is less then 45%.

• Remember, a p-value of 0.0244 would allow us to reject the null hypothesis at the 5% level, but not at the 1% level. A sample $\hat{p}$ value of 0.419 or less would occur only about 2.44% of the time, purely due to chance, if the true proportion of all adults who drank diet soda really were 45%. This gives us evidence to reject the null hypothesis at the 5% level.

8.3 Two-Sample z-Interval for Difference Between Two Proportions

- We are sometimes interested in comparing the proportion of successes between two groups. For example, we might be interested in knowing the difference in the proportion of males and females who text while driving. Or, we might be interested in the difference between the portion of college and high-school students who use laptops on a regular basis in their classes.

- We use the statistic $\hat{p}_1 - \hat{p}_2$ to estimate the true difference in the population proportions, $p_1 - p_2$. Remember that $\hat{p}_1 - \hat{p}_2$ is an unbiased estimator of $p_1 - p_2$.

- The standard deviation of the sampling distribution of $\hat{p}_1 - \hat{p}_2$ is

$$\sqrt{\frac{p_1(1-p_1)}{n_1} + \frac{p_2(1-p_2)}{n_2}}.$$

When dealing with a confidence interval, the values of p_1 and p_2 are unknown. For this reason, we use the standard error of the statistic $\hat{p}_1 - \hat{p}_2$:

$$SE = \sqrt{\frac{\hat{p}_1(1-\hat{p}_1)}{n_1} + \frac{\hat{p}_2(1-\hat{p}_2)}{n_2}}$$

- The formula for the confidence interval for comparing two proportions is:

$$\left(\hat{p}_1 - \hat{p}_2\right) \pm z^* \sqrt{\frac{\hat{p}_1(1-\hat{p}_1)}{n_1} + \frac{\hat{p}_2(1-\hat{p}_2)}{n_2}}$$

• The assumptions and conditions for inference when working with two proportions are as follows:

Assumptions	Conditions
1. Samples are independent of each other	1. Is this reasonable?
2. Individuals in each sample are independent	2. Both samples are SRSs and $n < 10\%$ of population for both samples
3. Both samples are large enough	3. $np \geq 10$ and $n(1-p) \geq 10$ for both samples

• **Example 4:** Nolan hopes to determine the difference in the proportion of males and females who play video games at least 4 days per week. Two independent random samples of size 200 are obtained. Of the 200 girls surveyed, 158 play video games at least 4 days per week. Of the 200 boys surveyed, 176 play video games at least 4 days per week. Construct a 99% confidence interval to help answer Nolan's question.

Solution:

Step 1: We want to estimate $p_1 - p_2$.

$p_1 =$ *the proportion of boys who play video games at least 4 days per week*

$p_2 =$ *the proportion of girls who play video games at least 4 days per week*

$$\hat{p}_1 = \frac{176}{200} = 0.88 \; and \; \hat{p}_2 = \frac{158}{200} = 0.79$$

Assumptions and conditions that verify:

1. **Samples are independent of each other.** We are given that the samples are independent of one another.

2. **Individuals in each sample are independent.** Both samples are given to be random, and we can safely assume that there are more than 2000 boys and 2000 girls in the population ($10n < N$ for both samples).

3. **Both samples are large enough:** $200(0.88) \geq 10$ and $200(0.12) \geq 10$.
   $$200(0.79) \geq 10 \text{ and } 200(0.21) \geq 10.$$

Step 2: The assumptions and conditions are met; we are safe to construct a two-proportion z-interval.

$$\left(\hat{p}_1 - \hat{p}_2\right) \pm z^* \sqrt{\frac{\hat{p}_1(1-\hat{p}_1)}{n_1} + \frac{\hat{p}_2(1-\hat{p}_2)}{n_2}}$$

$$\left(0.88 - 0.79\right) \pm 2.576 * \sqrt{\frac{0.88(1-0.88)}{200} + \frac{0.79(1-0.79)}{200}}$$

$$\left(-0.0049, 0.1849\right)$$

Step 3: We are 99% confident that the true difference in the proportion of boys and girls who play video games at least 4 days per week is between –.49% and 18.49%.

• Keep in mind that the interval contains zero. This means that zero is a plausible value for the difference in the proportion of boys and girls who play video games at least 4 days per week. Zero is contained in the interval, and we are 99% confident that the true difference is captured in the confidence interval that we have obtained. Note that a 95% confidence interval (0.00429, 0.17571) does not contain zero.

8.4 Two-Sample z-Test for Difference Between Two Proportions

• We are sometimes interested in knowing whether or not two population proportions are really different from one another. Remember that when sampling, we always encounter sampling variability. When we find the sample proportions, we want to determine if there really is a difference between the population proportions or if the difference between the obtained sample proportions is purely due to chance.

• The null hypothesis for a two-proportion z-test is $H_0 : p_1 = p_2$. The alternative hypothesis can be one- or two-sided. To conduct the hypothesis test, we need to standardize the test statistic, $\hat{p}_1 - \hat{p}_2$. If the null hypothesis is true, then the observations from each sample actually belong to a singe population. Therefore, instead of estimating each sample proportion separately, we use the **pooled sample proportion.** To find the *pooled sample proportion*, we use:

$$\hat{p} = \frac{combined\ successes\ in\ both\ samples}{combined\ observations\ in\ both\ samples}$$

• Once the pooled sample proportion is calculated, we can then find the standard error of the sampling distribution. Recall that for the two-proportion z-interval, we used:

$$SE = \sqrt{\frac{\hat{p}_1(1-\hat{p}_1)}{n_1} + \frac{\hat{p}_2(1-\hat{p}_2)}{n_2}}$$

Replacing each sample proportion with the pooled proportion, we obtain:

$$SE = \sqrt{\frac{\hat{p}(1-\hat{p})}{n_1} + \frac{\hat{p}(1-\hat{p})}{n_2}}$$

We can simplify the standard error to obtain:

$$SE = \sqrt{\hat{p}\left(1-\hat{p}\right)\left(\frac{1}{n_1} + \frac{1}{n_2}\right)}$$

• The assumptions and conditions for inference for a two-proportion z-test are the same as those for a two-proportion z-interval. Provided the assumptions and conditions for inference are met, we can use the test statistic:

$$z = \frac{\hat{p}_1 - \hat{p}_2}{\sqrt{\hat{p}\left(1-\hat{p}\right)\left(\frac{1}{n_1} + \frac{1}{n_2}\right)}}$$

• **Example 5:** A local college is interested in knowing if the proportion of high school students taking a fourth year of math is greater for students in suburban schools than in rural schools. To help answer this question, the college obtains a random sample of 500 senior students from suburban school districts and a random sample of 450 senior students from rural school districts across the country. The survey reveals that, of the 500 suburban students, 323 are taking a fourth year of math while only 279 of the 450 rural students are taking a fourth year of math. Is there evidence to suggest that the proportion of seniors taking a fourth year of math in suburban school districts is greater than the proportion of seniors taking a fourth year of math in rural school districts at the 5% level?

Solution:

Step 1:

p_1 = *proportion of seniors taking a fourth year of math*
     in suburban schools

p_2 = *proportion of seniors taking a fourth year of math*
     in rural schools

$$\hat{p}_1 = \frac{323}{500} = 0.646$$

$$\hat{p}_2 = \frac{279}{450} = 0.62$$

$H_0 : p_1 = p_2$

$H_a : p_1 > p_2$

Assumptions and conditions that verify:

1. **Samples are independent of each other.** We can assume that the samples are independent of one another because they are taken from school districts in different areas.

2. **Individuals in each sample are independent.** Both samples are given to be random, and we can safely assume that there are more than 5000 seniors in suburban school districts and 4500 seniors in rural school districts ($10n < N$ for both samples).

3. **Both samples are large enough:** $500(0.646) \geq 10$ and $500(0.354) \geq 10$.
                              $450(0.62) \geq 10$ and $200(0.38) \geq 10$.

Step 2: Because the assumptions and conditions for inference have been met, we should be safe to perform a two-sample z-test.

$$z = \frac{\hat{p}_1 - \hat{p}_2}{\sqrt{\hat{p}(1-\hat{p})\left(\dfrac{1}{n_1} + \dfrac{1}{n_2}\right)}}$$

$$pooled\ proportion\ \hat{p} = \frac{323 + 279}{500 + 450} \approx 0.6337$$

$$z = \frac{0.646 - 0.62}{0.6337(1-0.6337)\left(\dfrac{1}{500} + \dfrac{1}{450}\right)}$$

$$z \approx 0.8305 \quad p \approx 0.2031$$

Step 3: With a p-value of 0.2031, we fail to reject the null hypothesis at the 5% level. We conclude that the proportion of seniors taking a fourth year of math in suburban schools is not greater than the proportion of seniors taking a fourth year of math in rural schools.

Chapter

9

Inference for Related Variables: Chi-Square Distributions

9.1 The Chi-Square Statistic

• When our inference procedures involve categorical variables and our data are given in the form of counts, we turn to the **chi-square statistic** (χ^2). The chi-square statistic is actually a family of distributions and is always skewed to the right. Each of these distributions is classified by its *degrees of freedom*. Like the t-distributions, the distribution changes shape based on the degrees of freedom. As the degrees of freedom increase, the chi-square distributions become less skewed and become more symmetrical and more normal, as seen in Figure 9.1. All chi-square density curves start at zero on the x-axis, are single peaked, and approach the x-axis, asymptotically, as x increases (except when $df = 1$).

• The chi-square test statistic can be found using:

$$\chi^2 = \sum \frac{(O - E)^2}{E}$$ where O is the observed count and E is the expected count.

• We will discuss three types of tests involving the chi-square distributions. These include: *Chi-Square Test for Goodness of Fit, Chi-Square Test for Homogeneity of Populations, and the Chi-Square Test of Association/ Independence.* All three of these tests involve finding the same test statistic. We can find the p-value of each test by calculating the area under the chi-square distribution to the right of the test statistic. Remember that, like any density curve, the area under the chi-square distribution is equal to one.

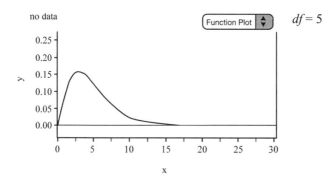

y = chiSquareDensity (x, 5, 1, 0)

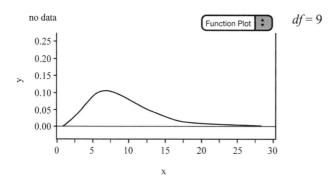

y = chiSquareDensity (x, 9, 1, 0)

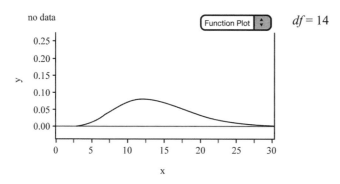

y = chiSquareDensity (x, 14, 1, 0)

Figure 9.1   Chi-square distributions with 5, 9, and 14 degrees of freedom.

• When performing chi-square tests of significance, we will use the familiar three-step format that we have used for all inference procedures. Again, there's nothing magical about the three steps; it's just a system you can use to ensure that you are always including the essentials of inference and that you are doing so in an organized fashion. The outline of the three steps is as follows:

1. Identify the appropriate type of chi-square test and verify that the assumptions and conditions for that test are met. State the null and alternative hypotheses in symbols or in words. Define any variables that you use.

2. Carry out the inference procedure. Do the math! Be sure to apply the correct formula and show the appropriate work.

3. Interpret the results in context of the problem.

9.2 Chi-Square Test for Goodness of Fit

• We sometimes want to examine the proportions in a single population. In this case, we turn to the *Chi-Square Test for Goodness of Fit*. You may have used or seen the chi-square test for goodness of fit in your biology class, for it is often used in the field of genetics. The goodness of fit test can be used by scientists to determine whether their hypothesized ratios are indeed correct. The null hypothesis in a goodness of fit test is that the actual population proportions are equal to the hypothesized values. The alternative hypothesis is that the actual population proportions are different from the hypothesized values.

• We can use the goodness of fit test to determine how well the observed counts match the expected counts. A classic example of the goodness of fit test is the M&M's candy activity. In this activity, we want to determine whether the M&M's candies are really manufactured in the proportions claimed by the manufacturer. This activity will help you understand when to use the goodness of fit test and how the goodness of fit test works. It could be implemented with any type of M&M's candies as long as you know the claimed proportions for each color. Skittles or any other type of candy or cereal could also be used provided you know the claimed proportions for each color. We will use this activity in Example 1 to perform a goodness of fit test.

• As with all inference, we must be sure to check the assumptions and conditions of the test. Following are the assumptions and conditions for the chi-square goodness of fit test:

Assumptions	Conditions
1. Data are in counts	1. Is this true?
2. Data are independent	2. SRS and <10% of population $(10n<N)$
3. Sample is large enough	3. All expected counts ≥ 5

• Once we have checked the assumptions and conditions for inference, we can calculate the chi-square test statistic to test the hypothesis of either a uniform distribution for the given categories or some specified distribution for each category. We can use the test statistic:

$$\chi^2 = \sum \frac{(O-E)^2}{E}$$

The chi-square statistic for a goodness of fit test has $n-1$ degrees of freedom, where n is the *number of categories* (not the sample size).

• We can calculate the p-value of the test by looking up the critical value with the correct degrees of freedom in the chi-square table of values or by using the graphing calculator χ^2 *cdf* command. We will discuss how to use the table of chi-square values in Example 1.

• **Example 1:** Mars Candy claims that plain M&M's candies are manufactured in the following proportions: 13% brown and red, 14% yellow, 24% blue, 20% orange, and 16% green. Using a 1.69-ounce bag of plain M&M's, test the manufacturer's claim at the 5% level of significance. For this example, we will use the following counts obtained from a 1.69-ounce bag of plain M&M's. We can find the expected number for each color by multiplying the total number of M&M's in the bag by the claimed proportion for each color. There were 56 M&M's in the bag. Figure 9.2 contains the observed counts as well as the expected counts for each of the six different colors. The expected counts for each color can be found by multiplying 56 (the total number of M&M's in the bag) by the corresponding claimed proportion for each color.

	Red	Yellow	Brown	Orange	Green	Blue
Observed	6	6	5	15	9	15
Expected	7.28	7.84	7.28	11.2	8.96	13.44

Figure 9.2   Observed and expected counts from a randomly selected 1.69-once bag of plain M&M's.

Solution:

Step 1: We will use a chi-square goodness of fit test to test the manufacturer's claim for the proportion of brown, red, yellow, orange, green, and blue M&M's.

H_0 : *The manufacturer's claim for the given proportions are correct That is:*

$p_{brown} = 0.13 \; p_{red} = 0.13 \; p_{yellow} = 0.14 \; p_{blue} = 0.24 \; p_{orange} = 0.20 \; p_{green} = 0.16$

H_a : *At least one of these proportions is incorrect*

Assumptions and conditions that verify:

1. **Data are in counts.** We can count the number of brown, red, yellow, orange, green, and blue M&M's in our sample.

2. **Data are independent.** We must consider our bag of M&M's to be a random sample. There are certainly more than 560 M&M's in the population of all plain M&M's ($10n < N$).

3. **Sample is large enough.** All expected counts in Figure 9.2 are greater than 5.

Step 2: With the assumptions and conditions of inference met, we should be safe to conduct a chi-square goodness of fit test. We find the test statistic using:

$$\chi^2 = \sum \frac{(O-E)^2}{E} \text{ with n–1 degrees of freedom}$$

$df = 5$ There are six categories (six colors).

$$\chi^2 = \frac{(6-7.28)^2}{7.28} + \frac{(6-7.84)^2}{7.84} + \frac{(5-7.28)^2}{7.28} + \frac{(15-11.2)^2}{11.2} + \frac{(9-8.96)^2}{8.96} + \frac{(15-13.44)^2}{13.44} \approx 2.8415$$

$p \approx 0.7244$

Step 3: With a p-value of approximately 0.7244, we fail to reject the null hypothesis at the 5% level of significance. We conclude that the proportions of colors of M&M's candies are not different from the proportions claimed by the manufacturer.

• In step 2, we obtained a p-value of 0.7244. We can interpret the p-value to mean the following: If repeated samples were taken (that is, many different bags of M&M's), we would anticipate observed counts as different or more different from the expected counts as we have obtained about 72% of the time, given that the claimed proportions by the manufacturer are really true. In other words, it's quite likely that the difference we are observing between the observed counts and expected counts is really just due to chance (sampling variability).

• How do we obtain the p-value of 0.7244? There are two methods, as mentioned earlier in this chapter. The first method is to use the χ^2 table of values to approximate the p-value. Remember that we are testing the manufacturer's claim at the 5% level. To use the table, we need to determine the **critical value.** The critical value is based, in part, by the level of significance at which we want to test our claim, and in part to the degrees of freedom. Using the χ^2 table of values, we can locate the critical value by cross-referencing 0.05 at the top of the table with 5 degrees of freedom. The corresponding critical value is 11.07. If we obtain a χ^2 test statistic greater than the critical value of 11.07, then we know that the corresponding p-value would be less than 0.05, which would lead us to reject the null hypothesis. Because our χ^2 value was only 2.8415, which is less than 11.07, we know the p-value is greater than 0.05. In fact, if we examine the table a little more closely, we can see that the smallest critical value for 5 degrees of freedom is 6.63. Our χ^2 value of 2.8415 is smaller than 6.63. We can therefore conclude that the p-value for our test will be greater than .25. Thus, we fail to reject the null hypothesis at the 5% level of significance. Using the critical value to estimate the p-value can also be used when working with t-distributions. Typically, we use our calculators to find the p-value by performing the appropriate test command.

• The second and most common way of finding the p-value for a chi-square goodness of fit test is to use the graphing calculator. Some graphing calculators have the goodness of fit test built into them. This makes it easy to find both the test statistic and the p-value. Some TI calculators have this test; others do not. Because some do not, we will briefly describe how to obtain the p-value for the goodness of fit test when the test is not built into the calculator.

- The TI-83 and TI-84 are both capable of creating lists. Place the observed values in List 1 and the expected values in List 2. Define List 3 to be

$$\frac{\left(L_1 - L_2\right)^2}{L_2}$$

We can then use the sum command, which is found under *2nd STAT (LIST), MATH*. The value obtained for the sum of List 3 is the test statistic. We then use the command *2nd VARS (DISTR)* and use the χ^2 command to determine the p-value.

9.3 Chi-Square Test for Homogeneity of Populations

- In Chapter 8, we discussed how to compare two proportions from two different groups using two-proportion z-procedures. We sometimes need to compare proportions across multiple groups. When we want to know if category proportions are the same for each group, we use the *Chi-Square Test for Homogeneity*. The data typically appear in two-way tables, as there are sometimes several categories. The chi-square test of homogeneity of populations eliminates the problem of comparing proportion 1 to proportion 2, proportion 1 to proportion 3, proportion 2 to proportion 3, and so on, as would be the case using multiple z-proportions.

• Although we are trying to determine whether the proportions for multiple populations are the same, it's important to remember that we are still working with counts. The expected counts for a chi-square test of homogeneity are *not* found in the same manner as they are in a goodness of fit test. To find the expected counts for a chi-square test of homogeneity, we use the following:

$$Expected\ cell\ count = \frac{(row\ total)(column\ total)}{(overall\ total)}$$

• The degrees of freedom are also calculated differently in a chi-square test of homogeneity than they are for a goodness of fit test. To find the degrees of freedom for a chi-square test of homogeneity, we use the following:

$$Degrees\ of\ freedom = (\#\ of\ rows - 1)(\#\ of\ columns - 1) = (r - 1)(c - 1)$$

• The null hypothesis for a chi-square test of homogeneity is that the distribution (proportion) of the counts for each group is the same. The alternative hypothesis is that the distribution for the counts for each group is not the same. We can write the null and alternative hypotheses in words or symbols.

• Because we are working with observed and expected counts, the chi-square test for homogeneity uses the same test statistic as the goodness of fit test.

$$\chi^2 = \sum \frac{(O - E)^2}{E}$$

• As is the case for all inference procedures, we must always check the assumptions and conditions. The assumptions and conditions for a chi-square test of homogeneity are:

Assumptions	Conditions
1. Data are in counts	1. Is this true?
2. Data in each sample are independent	2. SRS's and each sample <10% of population ($10n<N$)
3. Samples are large enough	3. All expected counts ≥ 5

• Consider the following hypothetical example involving the comparison of three proportions from three different populations.

• **Example 2:** A group of physicians specializing in weight loss is interested in knowing whether appetite suppressants are effective in helping people lose weight. They are curious to know if they should recommend regular exercise, appetite suppressants, or both to their patients. Suppose that a controlled experiment were conducted yielding the following results (see Figure 9.3). We will consider the proportion of those who lose at least 10 lbs. in a four-week period of time to be a success.

Treatment	Success	Failure	Total
Exercise Only	96 (94.576)	144 (145.42)	240
Drug Only	89 (94.576)	151 (145.42)	240
Exercise & Drug	103 (96.547)	142 (148.45)	245
Exercise & Placebo	95 (90.636)	135 (139.36)	230
Placebo Only	82 (88.665)	143 (136.33)	225
Total	465	715	1180

Figure 9.3 Homogeneity.

Solution:

Step 1: We want to compare the proportions of patients who lost at least 10 lbs. in a four-week period in the populations of patients who used exercise only (p_1), did not exercise but took an appetite suppressant (p_2), exercised and took the suppressant (p_3), exercised and took a placebo (p_4), and took a placebo only (p_5). We will use a chi-square test for homogeneity of populations.

$H_0 : p_1 = p_2 = p_3 = p_4 = p_5$

$H_a :$ *Not all five proportions are equal*

Assumption and conditions that verify:

Data are in counts. All sample data given in the two-way table are in counts.

Data are independent. We are given that the patients were randomly assigned to the treatment groups. We are safe to assume that the population of people for each group is easily 10 times the sample size ($10n<N$).

Sample is large enough. All expected counts in Figure 9.3 are greater than 5.

Step 2: With the assumptions and conditions met, we will conduct a chi-square test for homogeneity of populations. We can find the test statistic using:

$$\chi^2 = \sum \frac{(O-E)^2}{E}$$

$$\chi^2 = \frac{(96-94.576)^2}{94.576} + \frac{(144-145.42)^2}{145.42} + \frac{(89-94.576)^2}{94.576} + \frac{(151-145.42)^2}{145.42} + \frac{(103-96.547)^2}{96.547} +$$
$$\frac{(142-148.45)^2}{148.45} + \frac{(95-90.636)^2}{90.636} + \frac{(135-139.36)^2}{139.36} + \frac{(82-88.665)^2}{88.665} + \frac{(143-136.33)^2}{136.33}$$

$$\chi^2 \approx 2.4636 \quad p \approx 0.6512$$

Step 3: With a p-value of 0.6512, we fail to reject the null hypothesis. We conclude that there is not a difference in the proportions of patients who would lose at least 10 lbs. in a four-week period in the populations of patients who: exercise only (p_1), do not exercise but take an appetite suppressant (p_2), exercise and take the suppressant (p_3), exercise and take a placebo (p_4), and take a placebo only (p_5). The appetite suppressant does not appear to help patients lose weight.

9.4 Chi-Square Test for Independence/Association

• We use a *Chi-Square Test for Independence/Association* to determine whether there is an association between two categorical variables in a *single* population. As with the chi-square test for homogeneity, the data are usually given in two-way tables. When testing for independence/association, the two-way tables are called **contingency tables** because we are classifying individuals into *two* categorical variables.

• When do you use a chi-square test of homogeneity, and when do you use a chi-square test for independence/association? In order to differentiate between the two types of tests, you need to think about the design of the study.

• Remember that in a test of independence/association, there is a *single* sample from a *single* population. The individuals within the samples are classified according to two categorical variables. The chi-square test for homogeneity, on the other hand, takes only one sample from each of the populations of interest. Each individual from the sample is categorized based on a single variable. Thus, the null and alternative hypotheses differ depending on how the study was designed.

• The null hypothesis for a chi-square test of association/independence is that there is no relationship between the two categorical variables of interest. The alternative hypothesis is that there is a relationship between the two categorical variables of interest. We typically write the null and alternative in one of the following two ways:

H_0 : *The two categorical variables are independent*

H_a : *The two categorical variables are not independent*

or

H_0 : *There is no association between the categorical variables*

H_a : *There is an association between the categorical variables*

• We will use the same three-step procedure we have used for all inferences thus far, including the assumptions and conditions for the chi-square test for independence/association. The assumptions and conditions for this test are:

Assumptions	Conditions
1. Data are in counts	1. Is this true?
2. Data are independent	2. SRS and <10% of population ($10n<N$)
3. Sample is large enough	3. All expected counts ≥ 5

• Since the data are in counts, we continue to use the same chi-square test statistic:

$$\chi^2 = \sum \frac{(O-E)^2}{E}$$

• **Example 3:** You wish to evaluate the association between a person's gender and attitude toward spending money on public education. You obtain a random sample from your community and construct the contingency table shown in Figure 9.4.

Opinion	Female	Male	Total
Spend Less	40 (32.828)	28 (35.172)	68
Spend Same	14 (14.483)	16 (15.517)	30
Spend More	16 (22.69)	31 (24.31)	47
Total	70	75	145

Figure 9.4 Association/Independence.

Is there a relationship between gender and attitudes toward educational spending? Conduct an appropriate test to answer this question.

Solution:

Step 1: We are interested in knowing whether there is an association between a person's gender and attitude toward spending money on public education. We have obtained a single sample from a single population, so we will conduct a chi-square test for association/independence. The null and alternative hypotheses are:

H_0 : *There is no association between gender and attitudes toward educational spending*

H_0 : *There is an association between gender and attitudes toward educational spending*

We can check the appropriate assumptions and conditions.

Assumption and conditions that verify:

Data are in counts. All sample data given in the two-way table are in counts.

Data are independent. Our sample is random. We are safe to assume that the population of people is 10 times the sample size ($10n$<N).

Sample is large enough. All expected counts in Figure 9.4 are greater than 5.

Step 2: We have verified the conditions for inference for a chi-square test of association/independence. We are safe to find the chi-square test statistic:

$$\chi^2 = \sum \frac{(O-E)^2}{E}$$

$$\chi^2 = \frac{(40-32.828)^2}{32.828} + \frac{(28-35.172)^2}{35.172} + \frac{(14-14.483)^2}{14.483} + \frac{(16-15.517)^2}{15.517} + \frac{(16-22.69)^2}{22.69} + \frac{(31-24.31)^2}{24.31}$$

$$\chi^2 \approx 6.8740 \qquad p \approx 0.0322$$

Step 3: With a p-value of 0.0322, we reject the null hypothesis. There appears to be significant evidence (small p-value) to suggest that there is an association between a person's gender and attitudes toward spending money on public education.

Chapter

10

Inference for Regression

10.1 The Regression Model

• **Inference for regression** is the final type of inference discussed in AP Statistics. We use inference for regression when dealing with *two quantitative variables*. In Chapter 2, we discussed how to model data by using linear, exponential, and power functions. This chapter focuses on how to create confidence intervals for the slope of the least-squares regression line, $\hat{y} = a + bx$. You will also learn how to perform a hypothesis test for the slope of a linear relationship.

• In Chapter 2, we created the least-squares regression line, $\hat{y} = a + bx$ from the sample data collected from the population. As such, the slope b, and the y-intercept a, are statistics, not parameters. Remember that, due to sampling variability, the statistics a and b would probably take on different values if we took multiple samples. In other words, if another sample were taken, different data points would produce a different least-squares regression line and, consequently, different values of a and b. Recall that the least-squares regression line is formed by making the sum of the squares of the residuals as small as possible. Also remember that a residual is the difference in the observed value and the predicted value of y.

$$residual = observed\ y - predicted\ y = y - \hat{y}$$

• Remember that we use the least-squares regression line to make predictions of the response variable, y, based on the explanatory variable, x. We will use the statistics a and b as *unbiased estimates* of the true slope and y-intercept, which are the unknown parameters, α and β. The mean, μ_y, of all responses has a linear relationship with x that represents the true regression line where:

$$\mu_y = \alpha + \beta x$$

• Now that we've discussed how to estimate the slope and y-intercept of the true regression line, it's time to discuss the third parameter of interest in inference for regression, the standard deviation, σ. The standard deviation, σ, is used to measure the variability of the response variable y about the true regression line. Remember that the predicted values, $\hat{y}$, are on the regression line. The observed values of y vary about the regression line for any given value of x.

• If we are given n data points, there will be n residuals. Since σ is the standard deviation of the responses about the true regression line, we estimate σ by using the standard deviation of the residuals of the sample data. Because we are using an estimate for σ, we call the standard deviation the *standard error* as we have done in previous chapters. We refer to the standard deviation of the response variable as s.

• The **standard error about the regression line** is:

$$s = \sqrt{\frac{1}{n-2}\sum residual^2}$$

or

$$s = \sqrt{\frac{1}{n-2}\sum (y - \hat{y})^2}$$

• Typically, we use our calculators to find the value of s. In some problems, the value of s will be given.

• Notice that the formula for the standard deviation involves averaging the squared residuals (deviations) from the line. When we find the average of these squared deviations, we are dividing by $n - 2$. Because we are working with two variables, we use $n - 2$ degrees of freedom instead of $n - 1$ degrees of freedom. Be sure to state the degrees of freedom when performing inference for regression.

• The next two sections of this chapter will outline the steps of inference for a confidence interval for the slope β as well as a hypothesis test. To be consistent, we will outline the procedures using the familiar three-step process that we have utilized in previous chapters. Recall the following steps:

1. Identify the parameter of interest, choose the appropriate inference procedure, and verify that the assumptions and conditions for that procedure are met.
2. Carry out the inference procedure. Do the math! Be sure to apply the correct formula.
3. Interpret the results in context of the problem.

• As always, we must check the assumptions and conditions for inference. The following are the assumptions and conditions necessary for inference for regression:

Assumptions	Conditions
1. Relationship has linear form	1. Scatterplot is approximately linear
2. Residuals are independent	2. Residual plot does not have a definite pattern
3. Variability of residuals is constant	3. Residual plot has even spread
4. Residuals are approximately normal	4. Graph of residuals is approximately symmetrical and unimodal, or normal probability plot is approximately linear

10.2 Confidence Intervals for the Slope β

- Of the three parameters discussed in this chapter, the slope is the primary focus of inference when it comes to inference for regression in AP Statistics. Remember that the slope is a rate of change. It is the average rate of change in the response variable, y, as the explanatory variable, x, increases by one unit. Because the slope of the true regression line is unknown, we often want to estimate it using a confidence interval.

- When the conditions for regression inference are met, the estimated regression slope follows a t-distribution with $n - 2$ degrees of freedom. When finding a confidence interval for the slope of the least squares regression line, we use the familiar form: estimate $\pm$ margin of error. The formula is:

$b \pm t^* SE_b$ where SE_b is the standard error of the slope. We can find SE_b by using the following formula:

$$SE_b = \frac{s}{\sqrt{\sum \left(x - \bar{x}\right)^2}}$$

- It is very unlikely that you would have to use this formula on the AP* Exam, as it tends to be a tedious calculation, even with a calculator. Regression software, like Minitab, is capable of giving the needed values.

- **Example 1:** A group of teachers is interested in knowing whether a relationship exists between the average number of hours studied per week and high school cumulative grade point average (G.P.A.). The teachers obtain a random sample of students and determine the average number of hours each student studies along with the student's cumulative high

school G.P.A. Construct a 95% confidence interval for the true slope of
the regression line to help answer the teachers' question. Figure 10.1
presents a data table containing the average number of hours studied
per week and the corresponding G.P.A for the 20 high-school students
in the sample, along with a scatterplot of the data.

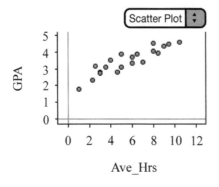

	Ave_Hrs	GPA
1	10.5	4.571
2	3.0	2.800
3	6.5	3.888
4	8.0	4.055
5	8.5	3.920
6	2.5	3.134
7	9.5	4.445
8	1.0	1.777
9	4.6	2.770
10	3.5	3.112
11	2.2	2.308
12	6.0	3.665
13	8.0	4.500
14	6.0	3.333
15	5.0	3.100
16	3.0	2.723
17	5.0	3.888
18	4.0	3.500
19	9.0	4.334
20	7.0	3.388

Figure 10.1 A scatterplot of the data appears roughly linear with no
apparent outliers.

Solution:

Step 1: We want to estimate β, the true slope of the regression line for the linear relationship between the average amount of time spent studying per week and the cumulative G.P.A. As always, we will check the assumptions and conditions necessary for inference.

Assumptions and conditions that verify:

1. **Relationship has a linear form.** The scatterplot in Figure 10.1 appears to be roughly linear.

2. **Residuals are independent.** The residual plot in Figure 10.2 shows no obvious pattern.

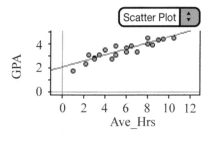

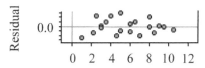

$$GPA = 0.256Ave\_Hrs + 2; r^2 = 0.8$$

Figure 10.2 The residual plot shows no definite pattern and appears to have even spread.

3. **Variability of residuals is constant.** The residual plot in Figure 10.2 appears to have even spread.

4. **Residuals are approximately normal.** The normal probability plot (Figure 10.3) is approximately linear.

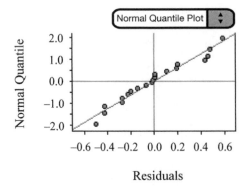

Residuals

Figure 10.3   The normal probability plot of the residuals
appears to be roughly linear.

Step 2: With the assumptions and conditions for regression inference
met, we are safe to construct a 95% confidence interval for the slope of
the true regression line.

$b \pm t^* \, SE_b$ with $n - 2$ degrees of freedom

$$0.2563 \pm 2.101 \frac{0.3277}{11.7604}$$

$(0.1975, 0.3151)$

Step 3: We are 95% confident that the slope of the true regression line is
between 0.1975 and 0.3151.

• Note that 0 is not included in the confidence interval. This implies that
the slope of the regression line is not equal to zero. This means that there
does appear to be a relationship between the average amount of time
spent studying per week and a student's cumulative grade point average.

- In order to find SE_b and s, run the linear regression t-test on your graphing calculator, which gives s. You can then find $\sqrt{\sum\left(x-\overline{x}\right)^2}$ by defining a list to be $\left(x-\overline{x}\right)^2$. For example, you could define $L_3 = (L_1 - 5.64)^2$ and then use the sum and square root functions on your calculator. Again, in most cases, SE_b and s will be given as computer output, and you'll just have to substitute them into the formula.

10.3 Hypothesis Testing for the Slope β

- We are now ready to discuss hypothesis testing for the slope β. If there is a relationship between the two quantitative variables of interest, the slope of the regression equation should be significantly different from zero.

- The null and alternative hypotheses for such a test are as follows:

$H_0 : \beta = 0$

$H_a : \beta \neq 0$ (or < 0 or > 0)

As usual, the alternative hypothesis can be one-sided or two-sided.

- The test-statistic associated with a test for the slope β is:

$$t = \frac{b-\beta}{SE_b} \quad \text{where} \quad SE_b = \frac{s}{\sqrt{\sum\left(x-\overline{x}\right)^2}}$$

- The assumptions and conditions for testing the slope β are the same as those for a confidence interval.

• **Example 2:** Is there reason to believe that a relationship exists between heights of fathers and the heights of their sons? To answer this question, use the following data in Figure 10.4, obtained from a random sample of 10 men and their sons. Test at the 5% level of significance.

	Father_Height	Sons_Height
1	65	66
2	64	63
3	68	69
4	73	72
5	72	73
6	67	66
7	71	72
8	75	75
9	70	69
10	69	70

Figure 10.4   Heights of 10 randomly selected men and their sons.

Solution:

Step 1: We want to test the claim that there is a relationship between the heights of fathers and their sons. Let β = *true slope of the least-squares regression line.*

$H_0 : \beta = 0$

$H_a : \beta \neq 0$

Assumptions and conditions that verify:

1. **Relationship has a linear form.** The scatterplot in Figure 10.5 appears to be roughly linear.

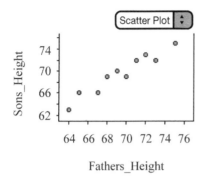

Figure 10.5 The scatterplot of father and son
heights appears roughly linear.

2. **Residuals are independent.** The residual plot in Figure 10.6 shows
no obvious pattern.

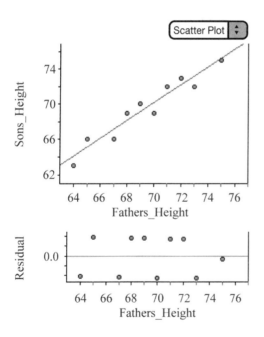

Sons_Height = 1.014Fathers_Height − 0.91; $r^2 = 0.93$

Figure 10.6 Residual plot shows no definite pattern
and appears to have even spread.

3. **Variability of residuals is constant.** The residual plot in Figure 10.6 appears to have even spread.

4. **Residuals are approximately normal.** The normal probability plot (Figure 10.7) is somewhat linear. Remember, if in doubt, you can always check a modified boxplot and look for skewness and outliers.

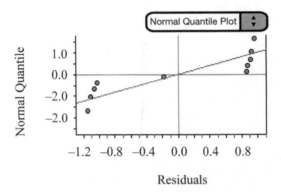

Figure 10.7　The normal probability plot of the residuals appears to be somewhat linear.

Step 2: With the assumptions and conditions for regression inference met, we are safe to proceed with the test for the true slope of the regression line.

$$t = \frac{b - \beta}{SE_b} \quad df = 8$$

$$t = \frac{1.0145 - 0}{0.1003}$$

$$t \approx 10.1193 \quad p \approx 7.7685 \times 10^{-6} \approx 0$$

Step 3: With a p-value of almost zero, we reject the null hypothesis at the 5% level. We conclude that the slope of the true regression line is different from zero and that there is a relationship between the heights of fathers and their sons.

• Note that the exact same procedures could be used to conduct a hypothesis test on *rho*, the correlation coefficient for the population.

• Statistics for regression are often given in the form of a Minitab printout. You will come across these printouts as you do "released" exam questions in both the multiple-choice and free-response sections. Be sure to work through a few "released" exam questions that include these printouts. Your instructor will likely provide you with such printouts as well. Remember that there are almost always some "extra" statistics that you do not need. Don't feel obligated to use all of the information from the printout. There may also be some statistics given that are not part of the AP Statistics curriculum. You can ignore what you don't need. You'll do great!

Appendix A

Tables

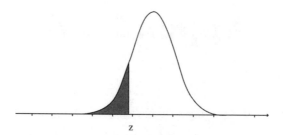

Table value for z is the probability to the left of z.

Table A: Standard Normal Probabilities

z	.00	.01	.02	.03	.04	.05	.06	.07	.08	.09
−3.4	.0003	.0003	.0003	.0003	.0003	.0003	.0003	.0003	.0003	.0002
−3.3	.0005	.0005	.0005	.0004	.0004	.0004	.0004	.0004	.0004	.0003
−3.2	.0007	.0007	.0006	.0006	.0006	.0006	.0006	.0005	.0005	.0005
−3.1	.0010	.0009	.0009	.0009	.0008	.0008	.0008	.0008	.0007	.0007
−3.0	.0013	.0013	.0013	.0012	.0012	.0011	.0011	.0011	.0010	.0010
−2.9	.0019	.0018	.0018	.0017	.0016	.0016	.0015	.0015	.0014	.0014
−2.8	.0026	.0025	.0024	.0023	.0023	.0022	.0021	.0021	.0020	.0019
−2.7	.0035	.0034	.0033	.0032	.0031	.0030	.0029	.0028	.0027	.0026
−2.6	.0047	.0045	.0044	.0043	.0041	.0040	.0039	.0038	.0037	.0036
−2.5	.0062	.0060	.0059	.0057	.0055	.0054	.0052	.0051	.0049	.0048
−2.4	.0082	.0080	.0078	.0075	.0073	.0071	.0069	.0068	.0066	.0064
−2.3	.0107	.0104	.0102	.0099	.0096	.0094	.0091	.0089	.0087	.0084
−2.2	.0139	.0136	.0132	.0129	.0125	.0122	.0119	.0116	.0113	.0110
−2.1	.0179	.0174	.0170	.0166	.0162	.0158	.0154	.0150	.0146	.0143
−2.0	.0228	.0222	.0217	.0212	.0207	.0202	.0197	.0192	.0188	.0183
−1.9	.0287	.0281	.0274	.0268	.0262	.0256	.0250	.0244	.0239	.0233
−1.8	.0359	.0351	.0344	.0336	.0329	.0322	.0314	.0307	.0301	.0294
−1.7	.0446	.0436	.0427	.0418	.0409	.0401	.0392	.0384	.0375	.0367
−1.6	.0548	.0537	.0526	.0516	.0505	.0495	.0485	.0475	.0465	.0455
−1.5	.0668	.0655	.0643	.0630	.0618	.0606	.0594	.0582	.0571	.0559
−1.4	.0808	.0793	.0778	.0764	.0749	.0735	.0721	.0708	.0694	.0681
−1.3	.0968	.0951	.0934	.0918	.0901	.0885	.0869	.0853	.0838	.0823
−1.2	.1151	.1131	.1112	.1093	.1075	.1056	.1038	.1020	.1003	.0985
−1.1	.1357	.1335	.1314	.1292	.1271	.1251	.1230	.1210	.1190	.1170
−1.0	.1587	.1562	.1539	.1515	.1492	.1469	.1446	.1423	.1401	.1379
−0.9	.1841	.1814	.1788	.1762	.1736	.1711	.1685	.1660	.1635	.1611
−0.8	.2119	.2090	.2061	.2033	.2005	.1977	.1949	.1922	.1894	.1867
−0.7	.2420	.2389	.2358	.2327	.2296	.2266	.2236	.2206	.2177	.2148
−0.6	.2743	.2709	.2676	.2643	.2611	.2578	.2546	.2514	.2483	.2451
−0.5	.3085	.3050	.3015	.2981	.2946	.2912	.2877	.2843	.2810	.2776
−0.4	.3446	.3409	.3372	.3336	.3300	.3264	.3228	.3192	.3156	.3121
−0.3	.3821	.3783	.3745	.3707	.3669	.3632	.3594	.3557	.3520	.3483
−0.2	.4207	.4168	.4129	.4090	.4052	.4013	.3974	.3936	.3897	.3859
−0.1	.4602	.4562	.4522	.4483	.4443	.4404	.4364	.4325	.4286	.4247
−0.0	.5000	.4960	.4920	.4880	.4840	.4801	.4761	.4721	.4681	.4641

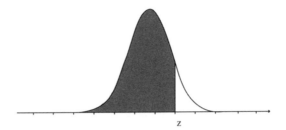

Table value for z is the probability to the left of z.

Table A: Standard Normal Probabilities *(continued)*

z	.00	.01	.02	.03	.04	.05	.06	.07	.08	.09
0.0	.5000	.5040	.5080	.5120	.5160	.5199	.5239	.5279	.5319	.5359
0.1	.5398	.5438	.5478	.5517	.5557	.5596	.5636	.5675	.5714	.5753
0.2	.5793	.5832	.5871	.5910	.5948	.5987	.6026	.6064	.6103	.6141
0.3	.6179	.6217	.6255	.6293	.6331	.6368	.6406	.6443	.6480	.6517
0.4	.6554	.6591	.6628	.6664	.6700	.6736	.6772	.6808	.6844	.6879
0.5	.6915	.6950	.6985	.7019	.7054	.7088	.7123	.7157	.7190	.7224
0.6	.7257	.7291	.7324	.7357	.7389	.7422	.7454	.7486	.7517	.7549
0.7	.7580	.7611	.7642	.7673	.7704	.7734	.7764	.7794	.7823	.7852
0.8	.7881	.7910	.7939	.7967	.7995	.8023	.8051	.8078	.8106	.8133
0.9	.8159	.8186	.8212	.8238	.8264	.8289	.8315	.8340	.8365	.8389
1.0	.8413	.8438	.8461	.8485	.8508	.8531	.8554	.8577	.8599	.8621
1.1	.8643	.8665	.8686	.8708	.8729	.8749	.8770	.8790	.8810	.8830
1.2	.8849	.8869	.8888	.8907	.8925	.8944	.8962	.8980	.8997	.9015
1.3	.9032	.9049	.9066	.9082	.9099	.9115	.9131	.9147	.9162	.9177
1.4	.9192	.9207	.9222	.9236	.9251	.9265	.9279	.9292	.9306	.9319
1.5	.9332	.9345	.9357	.9370	.9382	.9394	.9406	.9418	.9429	.9441
1.6	.9452	.9463	.9474	.9484	.9495	.9505	.9515	.9525	.9535	.9545
1.7	.9554	.9564	.9573	.9582	.9591	.9599	.9608	.9616	.9625	.9633
1.8	.9641	.9649	.9656	.9664	.9671	.9678	.9686	.9693	.9699	.9706
1.9	.9713	.9719	.9726	.9732	.9738	.9744	.9750	.9756	.9761	.9767
2.0	.9772	.9778	.9783	.9788	.9793	.9798	.9803	.9808	.9812	.9817
2.1	.9821	.9826	.9830	.9834	.9838	.9842	.9846	.9850	.9854	.9857
2.2	.9861	.9864	.9868	.9871	.9875	.9878	.9881	.9884	.9887	.9890
2.3	.9893	.9896	.9898	.9901	.9904	.9906	.9909	.9911	.9913	.9916
2.4	.9918	.9920	.9922	.9925	.9927	.9929	.9931	.9932	.9934	.9936
2.5	.9938	.9940	.9941	.9943	.9945	.9946	.9948	.9949	.9951	.9952
2.6	.9953	.9955	.9956	.9957	.9959	.9960	.9961	.9962	.9963	.9964
2.7	.9965	.9966	.9967	.9968	.9969	.9970	.9971	.9972	.9973	.9974
2.8	.9974	.9975	.9976	.9977	.9977	.9978	.9979	.9979	.9980	.9981
2.9	.9981	.9982	.9982	.9983	.9984	.9984	.9985	.9985	.9986	.9986
3.0	.9987	.9987	.9987	.9988	.9988	.9989	.9989	.9989	.9990	.9990
3.1	.9990	.9991	.9991	.9991	.9992	.9992	.9992	.9992	.9993	.9993
3.2	.9993	.9993	.9994	.9994	.9994	.9994	.9994	.9995	.9995	.9995
3.3	.9995	.9995	.9995	.9996	.9996	.9996	.9996	.9996	.9996	.9997
3.4	.9997	.9997	.9997	.9997	.9997	.9997	.9997	.9997	.9997	.9998

Table B: Random Digits

101	19223	95034	05756	28713	96409	12531	42544	82853
102	20826	04475	94351	33439	06175	03307	66233	22350
103	32071	50612	05952	30176	81471	85492	19221	57168
104	21199	86409	30003	05822	51816	41862	34876	76020
105	10455	84585	08091	48657	09879	32897	08853	62247
106	63443	55127	53478	41514	51995	07788	76625	79847
107	82739	57890	20807	47511	81676	55360	94072	02417
108	86824	94361	79090	63600	91936	51541	23963	88545
109	34683	84484	87891	83382	46973	93644	20814	86694
110	87605	13092	97004	12712	89090	34265	57676	08708
111	81486	69487	60513	09297	00412	71238	27649	39950
112	43857	27768	14703	90602	46070	36974	74691	79843
113	30596	91595	16816	21290	02139	39693	15210	67003
114	19762	60189	66177	65805	19248	68403	75439	53656
115	23720	26018	93503	26463	87803	71190	22160	87096
116	53696	38126	21384	38418	16164	69209	75415	76349
117	75230	18842	44098	40784	72352	42232	47847	85306
118	70494	19004	08513	16998	29286	06732	49386	06967
119	37863	76884	98089	18779	03689	77790	33927	95962
120	58433	44928	03929	27859	97915	52224	44127	18994
121	29690	37087	48463	02226	56877	22366	93941	74584
122	21123	86170	53411	30529	97469	24365	49980	95663
123	88890	58862	23433	47347	70805	22045	04481	06427
124	61999	42290	25045	80833	66492	48399	57551	87185
125	73846	21500	27562	03460	21513	50913	39017	46856
126	59321	17125	05234	13443	14507	39680	24430	49485
127	45415	07837	42078	93921	54639	76462	30303	38530
128	20766	40730	22023	50998	13032	37150	85029	55129
129	25942	85230	00761	84894	77771	60839	76961	98370
130	27900	40660	16853	06842	05640	21090	66672	35358
131	51643	57697	43214	90366	19817	43734	98927	15045
132	21683	72667	15479	11720	89142	59861	43415	48417
133	58974	43986	23733	01102	55638	46270	99271	40082
134	17761	08361	01221	72309	17597	48752	73073	03052
135	04460	63771	03141	70392	18954	44344	16167	01776
136	23626	34712	17032	17810	34740	26053	93307	16285
137	32790	17033	55420	97717	61006	64560	22480	07641
138	43736	71972	47283	53324	93486	10687	36572	22854
139	93745	92391	77040	95992	83135	30714	06719	59096
140	47160	09855	48906	85728	84229	80628	03316	57587
141	23867	62845	04461	57908	32363	29866	43069	00888
142	31972	82955	23581	26219	32396	98106	03259	13009
143	69479	02023	42240	58720	78179	51440	83402	34979
144	28722	34202	77501	93305	53698	55189	23017	53861
145	19687	12633	57857	95806	09931	02150	43163	58636
146	22232	22320	40740	78321	65478	77484	33012	69691

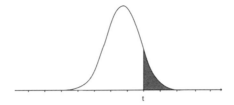

Table entry for *p* is the point t* with probability *p* to the right.

Table C: *t* Distribution Critical Values

df	.25	.20	.15	.10	.05	.025	.02	.01	.005	.0025	.001	.0005
						Upper tail probability *p*						
1	1.000	1.376	1.963	3.078	6.314	12.71	15.89	31.82	63.66	127.3	318.3	636.6
2	.816	1.061	1.386	1.886	2.920	4.303	4.849	6.965	9.925	14.09	22.33	31.60
3	.765	.978	1.250	1.638	2.353	3.182	3.482	4.541	5.841	7.453	10.21	12.92
4	.741	.941	1.190	1.533	2.132	2.776	2.999	3.747	4.604	5.598	7.173	8.610
5	.727	.920	1.156	1.476	2.015	2.571	2.757	3.365	4.032	4.773	5.893	6.869
6	.718	.906	1.134	1.440	1.943	2.447	2.612	3.143	3.707	4.317	5.208	5.959
7	.711	.896	1.119	1.415	1.895	2.365	2.517	2.998	3.499	4.029	4.785	5.408
8	.706	.889	1.108	1.397	1.860	2.306	2.449	2.896	3.355	3.833	4.501	5.041
9	.703	.883	1.100	1.383	1.833	2.262	2.398	2.821	3.250	3.690	4.297	4.781
10	.700	.879	1.093	1.372	1.812	2.228	2.359	2.764	3.169	3.581	4.144	4.587
11	.697	.876	1.088	1.363	1.796	2.201	2.328	2.718	3.106	3.497	4.025	4.437
12	.695	.873	1.083	1.356	1.782	2.179	2.303	2.681	3.055	3.428	3.930	4.318
13	.694	.870	1.079	1.350	1.771	2.160	2.282	2.650	3.012	3.372	3.852	4.221
14	.692	.868	1.076	1.345	1.761	2.145	2.264	2.624	2.977	3.326	3.787	4.140
15	.691	.866	1.074	1.341	1.753	2.131	2.249	2.602	2.947	3.286	3.733	4.073
16	.690	.865	1.071	1.337	1.746	2.120	2.235	2.583	2.921	3.252	3.686	4.015
17	.689	.863	1.069	1.333	1.740	2.110	2.224	2.567	2.898	3.222	3.646	3.965
18	.688	.862	1.067	1.330	1.734	2.101	2.214	2.552	2.878	3.197	3.611	3.922
19	.688	.861	1.066	1.328	1.729	2.093	2.205	2.539	2.861	3.174	3.579	3.883
20	.687	.860	1.064	1.325	1.725	2.086	2.197	2.528	2.845	3.153	3.552	3.850
21	.686	.859	1.063	1.323	1.721	2.080	2.189	2.518	2.831	3.135	3.527	3.819
22	.686	.858	1.061	1.321	1.717	2.074	2.183	2.508	2.819	3.119	3.505	3.792
23	.685	.858	1.060	1.319	1.714	2.069	2.177	2.500	2.807	3.104	3.485	3.768
24	.685	.857	1.059	1.318	1.711	2.064	2.172	2.492	2.797	3.091	3.467	3.745
25	.684	.856	1.058	1.316	1.708	2.060	2.167	2.485	2.787	3.078	3.450	3.725
26	.684	.856	1.058	1.315	1.706	2.056	2.162	2.479	2.779	3.067	3.435	3.707
27	.684	.855	1.057	1.314	1.703	2.052	2.158	2.473	2.771	3.057	3.421	3.690
28	.683	.855	1.056	1.313	1.701	2.048	2.154	2.467	2.763	3.047	3.408	3.674
29	.683	.854	1.055	1.311	1.699	2.045	2.150	2.462	2.756	3.038	3.396	3.659
30	.683	.854	1.055	1.310	1.697	2.042	2.147	2.457	2.750	3.030	3.385	3.646
40	.681	.851	1.050	1.303	1.684	2.021	2.123	2.423	2.704	2.971	3.307	3.551
50	.679	.849	1.047	1.299	1.676	2.009	2.109	2.403	2.678	2.937	3.261	3.496
60	.679	.848	1.045	1.296	1.671	2.000	2.099	2.390	2.660	2.915	3.232	3.460
80	.678	.846	1.043	1.292	1.664	1.990	2.088	2.374	2.639	2.887	3.195	3.416
100	.677	.845	1.042	1.290	1.660	1.984	2.081	2.364	2.626	2.871	3.174	3.390
1000	.675	.842	1.037	1.282	1.646	1.962	2.056	2.330	2.581	2.813	3.098	3.300
∞	.674	.841	1.036	1.282	1.645	1.960	2.054	2.326	2.576	2.807	3.091	3.291
	50%	60%	70%	80%	90%	95%	96%	98%	99%	99.5%	99.8%	99.9%
						Confidence level *C*						

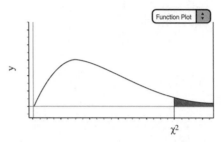

Table entry for p is the value χ^2 with probability p to the right.

Table D: $\chi 2$ Critical Values

df						Tail probability p					
	.25	.20	.15	.10	.05	.025	.02	.01	.005	.0025	.001
1	1.32	1.64	2.07	2.71	3.84	5.02	5.41	6.63	7.88	9.14	10.83
2	2.77	3.22	3.79	4.61	5.99	7.38	7.82	9.21	10.60	11.98	13.82
3	4.11	4.64	5.32	6.25	7.81	9.35	9.84	11.34	12.84	14.32	16.27
4	5.39	5.99	6.74	7.78	9.49	11.14	11.67	13.28	14.86	16.42	18.47
5	6.63	7.29	8.12	9.24	11.07	12.83	13.39	15.09	16.75	18.39	20.51
6	7.84	8.56	9.45	10.64	12.59	14.45	15.03	16.81	18.55	20.25	22.46
7	9.04	9.80	10.75	12.02	14.07	16.01	16.62	18.48	20.28	22.04	24.32
8	10.22	11.03	12.03	13.36	15.51	17.53	18.17	20.09	21.95	23.77	26.12
9	11.39	12.24	13.29	14.68	16.92	19.02	19.68	21.67	23.59	25.46	27.88
10	12.55	13.44	14.53	15.99	18.31	20.48	21.16	23.21	25.19	27.11	29.59
11	13.70	14.63	15.77	17.28	19.68	21.92	22.62	24.72	26.76	28.73	31.26
12	14.85	15.81	16.99	18.55	21.03	23.34	24.05	26.22	28.30	30.32	32.91
13	15.98	16.98	18.20	19.81	22.36	24.74	25.47	27.69	29.82	31.88	34.53
14	17.12	18.15	19.41	21.06	23.68	26.12	26.87	29.14	31.32	33.43	36.12
15	18.25	19.31	20.60	22.31	25.00	27.49	28.26	30.58	32.80	34.95	37.70
16	19.37	20.47	21.79	23.54	26.30	28.85	29.63	32.00	34.27	36.46	39.25
17	20.49	21.61	22.98	24.77	27.59	30.19	31.00	33.41	35.72	37.95	40.79
18	21.60	22.76	24.16	25.99	28.87	31.53	32.35	34.81	37.16	39.42	42.31
19	22.72	23.90	25.33	27.20	30.14	32.85	33.69	36.19	38.58	40.88	43.82
20	23.83	25.04	26.50	28.41	31.41	34.17	35.02	37.57	40.00	42.34	45.31
21	24.93	26.17	27.66	29.62	32.67	35.48	36.34	38.93	41.40	43.78	46.80
22	26.04	27.30	28.82	30.81	33.92	36.78	37.66	40.29	42.80	45.20	48.27
23	27.14	28.43	29.98	32.01	35.17	38.08	38.97	41.64	44.18	46.62	49.73
24	28.24	29.55	31.13	33.20	36.42	39.36	40.27	42.98	45.56	48.03	51.18
25	29.34	30.68	32.28	34.38	37.65	40.65	41.57	44.31	46.93	49.44	52.62
26	30.43	31.79	33.43	35.56	38.89	41.92	42.86	45.64	48.29	50.83	54.05
27	31.53	32.91	34.57	36.74	40.11	43.19	44.14	46.96	49.64	52.22	55.48
28	32.62	34.03	35.71	37.92	41.34	44.46	45.42	48.28	50.99	53.59	56.89
29	33.71	35.14	36.85	39.09	42.56	45.72	46.69	49.59	52.34	54.97	58.30
30	34.80	36.25	37.99	40.26	43.77	46.98	47.96	50.89	53.67	56.33	59.70
40	45.62	47.27	49.24	51.81	55.76	59.34	60.44	63.69	66.77	69.70	73.40
50	56.33	58.16	60.35	63.17	67.50	71.42	72.61	76.15	79.49	82.66	86.66
60	66.98	68.97	71.34	74.40	79.08	83.30	84.58	88.38	91.95	95.34	99.61
80	88.13	90.41	93.11	96.58	101.9	106.6	108.1	112.3	116.3	120.1	124.8
100	109.1	111.7	114.7	118.5	124.3	129.6	131.1	135.8	140.2	144.3	149.4

Appendix B

Formulas

Many formulas are given on the AP Statistics Exam. The first section of this appendix provides the formulas that you will be given during the AP Exam. The second section of this appendix is a summary of formulas not given on the AP Exam. You should be familiar with the formulas that will be given on the exam (but there's no need to memorize them). You will probably not use all of the formulas that are given. In fact, you may never use a few of them in the entire course. You should also know and understand all of the formulas in the second section of this appendix, because those will not be given to you on the exam. Knowing the formulas that will not be on the exam will put you at ease when solving problems and will let you focus on answering the questions and doing so in context. Also notice that some of the formulas given on the exam will help you remember the formulas that are not on the exam.

Formulas Given on the AP Exam

The following formulas for Descriptive Statistics, Probability, and Inferential Statistics are given on the AP Statistics Exam:

Descriptive Statistics

$$\bar{x} = \frac{\sum x_i}{n}$$

$$s_x = \sqrt{\frac{1}{n-1}\sum\left(x_i - \bar{x}\right)^2}$$

$$s_p = \sqrt{\frac{\left(n_1-1\right)s_1^2 + \left(n_2-1\right)s_2^2}{\left(n_1-1\right)+\left(n_2-1\right)}}$$

$$\hat{y} = b_0 + b_1 x$$

$$b_1 = \frac{\sum\left(x_i - \bar{x}\right)\left(y_i - \bar{y}\right)}{\sum\left(x_i - \bar{x}\right)^2}$$

$$b_0 = \bar{y} - b_1 \bar{x}$$

$$r = \frac{1}{n-1}\sum\left(\frac{x_i - \bar{x}}{s_x}\right)\left(\frac{y_i - \bar{y}}{s_y}\right)$$

$$b_1 = r\frac{s_y}{s_x}$$

$$s_{b_1} = \frac{\sqrt{\dfrac{\sum\left(y_i - \hat{y}_i\right)^2}{n-2}}}{\sqrt{\sum\left(x_i - \bar{x}\right)^2}}$$

Probability

$$P(A \cup B) = P(A) + P(B) - P(A \cap B)$$

$$P(A/B) = \frac{P(A \cap B)}{P(B)}$$

$$E(X) = \mu_x = \sum x_i P_i$$

$$Var(X) = \sigma_x^2 = \sum (x_i - \mu_x)^2 P_i$$

The following formulas are given and should be used if X has a binomial distribution with parameters n and p:

$$P(X = k) = \binom{n}{k} p^k (1-p)^{n-k}$$

$$\mu_x = np$$

$$\sigma_x = \sqrt{np(1-p)}$$

$$\mu_{\hat{p}} = p$$

$$\sigma_{\hat{p}} = \sqrt{\frac{p(1-p)}{n}}$$

If $\bar{x}$ is the mean of a random sample of size n from an infinite population with mean μ and standard deviation σ, then:

$$\mu_{\bar{x}} = \mu$$

$$\sigma_{\bar{x}} = \frac{\sigma}{\sqrt{n}}$$

Inferential Statistics

Standardized test statistic: $\dfrac{statistic - parameter}{std.\,deviation\,of\,statistic}$

Confidence interval: $statistic \pm (critical\,value) \bullet (std.\,deviation\,of\,statistic)$

Single-Sample

Statistic	**Standard Deviation of Statistic**
Sample mean	$\dfrac{\sigma}{\sqrt{n}}$
Sample proportion	$\sqrt{\dfrac{p(1-p)}{n}}$

Two-Sample

Difference of sample means $\quad\sqrt{\dfrac{\sigma_1^2}{n_1} + \dfrac{\sigma_2^2}{n_2}}$

Difference of sample means

(Special case when $\sigma_1 = \sigma_2$) $\quad \sigma\sqrt{\dfrac{1}{n_1} + \dfrac{1}{n_2}}$

Difference of sample proportions $\quad \sqrt{\dfrac{p_1(1-p_1)}{n_1} + \dfrac{p_2(1-p_2)}{n_2}}$

Difference of sample proportion

(Special case when $p_1 = p_2$) $\quad \sqrt{p(1-p)}\sqrt{\dfrac{1}{n_1} + \dfrac{1}{n_2}}$

Chi-square test statistic $\quad \chi^2 = \sum \dfrac{(observed - expected)^2}{expected}$

Formulas Not Given on the AP Exam

The following formulas are not given on the AP Exam. Be sure to know and understand how they work and how to apply them.

Normal Distribution

For normal distributions, use

z-score

$$z = \frac{x - \mu}{\sigma}$$

Probability

Probability that any two events A and B happen together:

$$P(A \cap B) = P(A) \cdot P(B \,/\, A)$$

Notice that this formula is given on the AP Exam, but in the following form:

$$P(A \,/\, B) = \frac{P(A \cap B)}{P(B)}$$

Probability that two *independent* events A and B happen together:

$$P(A \cap B) = P(A) \cdot P(B)$$

General addition rule for the union of two events:

$$P(A \cup B) = P(A) + P(B) - P(A \cap B)$$

Note that $\cup$ (union) means "or" and $\cap$ (intersection) means "and."

If the events A and B are disjoint (mutually exclusive):

$P(A \cup B) = P(A) + P(B)$

Rules for means and variances of random variables for fixed numbers *a* and *b*:

$\mu_z = a \pm b\mu_x$

$\sigma_z^2 = b^2\sigma_x^2$

$\sigma_z = b\sigma_x$

Addition rule for variances if X and Y are independent random variables:

$\sigma_{X+Y}^2 = \sigma_X^2 + \sigma_Y^2$

$\sigma_{X-Y}^2 = \sigma_X^2 + \sigma_Y^2$ (*This is not a typo! We always add variances!*)

Binomial distribution formula:

$$P = (X = k) = \binom{n}{k} p^k (1-p)^{n-k} \quad where:$$

$n = $ *number of trials*

$p = $ *probability of "success"*

$1 - p = $ *probability of "failure"*

$k = $ *number of successes in n trials*

$$\binom{n}{k} = \frac{n!}{(n-k)!k!}$$

Notice that the formula for the binomial distribution is given on the AP Exam. It's given in this section again to ensure that you understand the notation correctly.

Geometric distribution formulas

The probability that the first success is obtained on the nth observation:

$$P(X=n)=(1-p)^{n-1}\,p$$

Probability that it takes more than n trials to obtain the first success:

$$P(X>n)=(1-p)^{n}$$

Formulas for the mean and standard deviation of a geometric distribution:

$$\mu = \frac{1}{p} \quad \text{and} \quad \sigma = \sqrt{\frac{(1-p)}{p^2}}$$

Central limit theorem

The central limit theorem says that as the sample size increases, the mean of the sampling distribution of $\bar{x}$ approaches a normal distribution with mean, μ, and standard deviation,

$$\sigma_x = \frac{\sigma}{\sqrt{n}}.$$

Standard error

When using s to estimate σ, the standard deviation of the sampling distribution for means is $s_{\bar{x}} = \frac{s}{\sqrt{n}}$. When using s to estimate σ, the standard deviation of the sampling distribution is called the **standard error** of the sample mean, $\bar{x}$.

Inferential Statistics

The following formulas are used for inferential statistics:

Mean(s)

One sample t-interval:

$$\bar{x} \pm t^*_{n-1} \times \frac{s}{\sqrt{n}} \text{ with } n\text{–}1 \text{ degrees of freedom}$$

One sample t-test:

$$t = \frac{\bar{x} - \mu}{s / \sqrt{n}} \text{ with } n\text{–}1 \text{ degrees of freedom}$$

Two sample t-interval:

$$\left(\bar{x}_1 - \bar{x}\right) \pm t^* \times \sqrt{\frac{s_1^2}{n_1} + \frac{s_2^2}{n_2}}$$

The t^* value depends on the particular level

of confidence that you want and on the degrees of freedom (df).

Two sample t-test:

$$t = \frac{\left(\bar{x}_1 - \bar{x}\right) - \left(\mu_1 - \mu_2\right)}{\sqrt{\frac{s_1^2}{n_1} + \frac{s_2^2}{n_2}}}$$

Proportion(s)

Standard deviation of the sampling distribution of $\hat{p}$:

$$\sqrt{\frac{p(1-p)}{n}}$$

When dealing with confidence intervals, we do not know p. Because $\hat{p}$ is an unbiased estimator of p, we use $\hat{p}$ to estimate p. These two values should be close in value, provided that the sample is large enough. We can then use the standard error of $\hat{p}$:

$$SE = \sqrt{\frac{\hat{p}(1-\hat{p})}{n}}$$

One-proportion z-interval:

$$\hat{p} \pm z^* \sqrt{\frac{\hat{p}(1-\hat{p})}{n}}$$

One-proportion z-test:

$$z = \frac{(\hat{p} - p_0)}{\sqrt{\frac{p_0(1-p_0)}{n}}}$$

Note that p_0 is the value of the proportion in the null hypothesis.

Two-proportion z-interval:

$$\left(\hat{p}_1 - \hat{p}_2\right) \pm z^* \sqrt{\frac{\hat{p}_1(1-\hat{p}_1)}{n_1} + \frac{\hat{p}_2(1-\hat{p}_2)}{n_2}}$$

When dealing with a confidence interval, the values of p_1 and p_2 are unknown. For this reason, we use the standard error of the statistic $\hat{p}_1 - \hat{p}_2$:

$$SE = \sqrt{\frac{\hat{p}_1(1-\hat{p}_1)}{n_1} + \frac{\hat{p}_2(1-\hat{p}_2)}{n_2}}$$

Two-proportion z-test:

$$z = \frac{\hat{p}_1 - \hat{p}_2}{\hat{p}(1-\hat{p})\left(\dfrac{1}{n_1} + \dfrac{1}{n_2}\right)}$$

Use the pooled sample proportion when using a two-proportion test. To find the *pooled sample proportion*, we use:

$$\hat{p} = \frac{combined\ successes\ in\ both\ samples}{combined\ observations\ in\ both\ samples}$$

Regression

Mean of all responses of a linear relationship with x that represents the true regression line:

$$\mu_y = \alpha + \beta x$$

Standard error about the regression line:

$$s = \sqrt{\frac{1}{n-2}\sum residual^2}$$

or

$$s = \sqrt{\frac{1}{n-2}\sum(y-\hat{y})^2}$$

Confidence interval for the true slope (β) of the regression line:

$b \pm t * SE_b$

where SE_b is the standard error of the slope. We can find SE_b by using the following formula:

$$SE_b = \frac{s}{\sqrt{\sum (x - \bar{x})^2}}$$

- You probably will not need to find SE_b, as it is typically given on the exam.

Test for the slope β:

$$t = \frac{b - \beta}{SE_b}$$

where $SE_b = \dfrac{s}{\sqrt{\sum (x - \bar{x})^2}}$

Appendix C

Assumptions and
Conditions for Inference

The following are the assumptions and conditions essential for proper statistical inference. Be sure that you know and understand all of the assumptions and conditions for the various types of confidence intervals and tests. Make certain that you check them accordingly when conducting inference on the AP Exam. It will be assumed that you know that the assumptions and conditions should always be checked when doing inference. The test questions about inference will probably not remind you to check them.

One-sample t-interval or one-sample t-test

Assumptions	Conditions
1. Individuals are independent	1. SRS and <10% of population ($10n$<N)
2. Normal population assumption	2. One of the following: –Given a normal population –Graph of data is symmetric with no outliers –Sample is large enough ($n \geq 30$) that the sampling distribution of $\bar{x}$ is approximately normal

Remember that matched pairs are a one-sample t-procedure. Check the assumptions and conditions for a one-sample t-procedure when doing matched pairs. You should also be sure to state that the "data are matched," as this is an added assumption.

Two-sample t-interval or two-sample t-test

Assumptions	Conditions
1. Samples are independent of each other	1. Are they? Does this seem reasonable?
2. Individuals in each sample are independent	2. Both SRSs and both <10% population ($10n$<N for both samples)
3. Normal populations assumption	3. One of the following: –Given normal populations –Graph of data for both samples shows no outliers or strong skewness –Samples are both large ($n \geq 30$); therefore the sampling distribution of $\bar{x}_1 - \bar{x}_2$ is approximately normal

One-proportion z-interval or test

Assumptions	Conditions
1. Individuals are independent	1. SRS and $n < 10\%$ of population
2. Sample is large enough	2. $np \geq 10$ and $n(1-p) \geq 10$ Use $\hat{p}$ for C.I. and p_0 for Tests

Two-proportion z-interval or test

Assumptions	Conditions
1. Samples are independent of each other	1. Is this reasonable?
2. Individuals in each sample are independent	2. Both samples are SRSs and $n < 10\%$ of population for both samples
3. Both samples are large enough	3. $np \geq 10$ and $n(1-p) \geq 10$ for both samples

Chi-square goodness of fit (one variable from one sample)

Assumptions	Conditions
1. Data are in counts	1. Is this true?
2. Data are independent	2. SRS and <10% of population ($10n<N$)
3. Sample is large enough	3. All expected counts ≥ 5

Chi-square test for homogeneity (samples from many populations)

Assumptions	Conditions
1. Data are in counts	1. Is this true?
2. Data in each sample are independent	2. SRSs and each sample <10% of population ($10n<N$)
3. Samples are large enough	3. All expected counts ≥ 5

Chi-square test for independence/association
(one sample from one population classified on two variables)

Assumptions	Conditions
1. Data are in counts	1. Is this true?
2. Data are independent	2. SRS and <10% of population ($10n$<N)
3. Sample is large enough	3. All expected counts ≥ 5

Regression (t)

Assumptions	Conditions
1. Relationship has linear form	1. Scatterplot is approximately linear
2. Residuals are independent	2. Residual plot does not have a definite pattern
3. Variability of residuals is constant	3. Residual plot has even spread
4. Residuals are approximately normal	4. Graph of residuals is approximately symmetrical and unimodal, or normal probability plot is approximately linear

Glossary

Alternative hypothesis   Hypothesis that contains the value of the parameter to accept if the null hypothesis has been rejected.

Binomial distribution (Bernoulli trial)   Four conditions must be met in order for a distribution to be considered a binomial. These conditions are:

1. Each observation can be considered a "success" or "failure."
2. There must be a fixed number of trials or observations.
3. The observations must be independent.
4. The probability of success, which we call p, is the same from one trial to the next.

Block (blocking)   Used in experiments when it is believed that subjects or experimental units are different in some way that may affect the results of the experiment.

Categorical variable   Variable that places an individual into a category or group.

Census   Consists of all individuals in the entire population.

Center of the distribution   The mean and/or the median of the distribution is usually considered the center of the distribution.

Central limit theorem   The central limit theorem states that as the sample size increases, the mean of the sampling distribution of x approaches a normal distribution with mean, μ, and standard deviation,

$\sigma_x = \dfrac{\sigma}{\sqrt{n}}$.   This is true for any population, not just normal populations!

Chi-square statistic (x^2.)   The chi-square statistic is a family of distributions and is always skewed to the right. Each of these distributions is classified by its *degrees of freedom*. The chi-square test statistic can be found using:

$x^2 = \sum \dfrac{(O-E)^2}{E}$ where O is the observed count and E is the expected count.

Cluster sample   Similar to a stratified sample. In a cluster sample, however, the groups are heterogeneous, not homogeneous. That is, the groups will not necessarily differ from one another. Once the groups are determined, we can conduct an SRS within each group and form the entire sample from the results of each SRS.

Coefficient of determination   The r^2 value (coefficient of determination) is the amount of variability of y that can be explained or accounted for by the linear relationship of y on x. To find r^2, we simply square the r-value. Remember, even an r^2 value of 1 does not necessarily imply any cause-and-effect relationship!

Complement   The complement of event E is the event that E does not occur. The complement of event E is written as E^c.

Completely randomized experiment   Subjects or experimental units are randomly assigned to a treatment group. Completely randomized experiments can be used to compare any number of treatments. Groups of equal size should be used, if possible.

Conditional probability   Probability of an event if it is known that another event or condition has occurred or not occurred.

Confidence interval   The confidence interval gives a range of values that would be reasonable values for the parameter of interest, based on the statistic obtained from the sample.

Confounding variables   Variables, aside from the explanatory variable, that may affect the response variable.

Continuous random variable   Random variable that can take on values that comprise an interval of real numbers.

Control group   Used to help compare the various treatments. The control group can be used to help determine if the new treatment really does work or have a desired effect.

Convenience sample   Sample conducted due to the ease of data collection. Convenience samples typically contain bias.

Correlation coefficient   The correlation coefficient can be found by using the formula:

$$r = \frac{1}{n-1} \sum \left(\frac{x_i - \bar{x}}{s_x} \right)\left(\frac{y_i - \bar{y}}{s_y} \right)$$

Remember that correlation refers only to a linear relationship. Do not use the correlation coefficient to describe non-linear relationships! Correlation does not imply causation. Just because two variables are strongly associated or even correlated (linear) does not mean that changes in one variable are causing changes in another.

Counting principle or **multiplication principle**   The multiplication principle states that if you can do task 1 in m ways and you can do task 2 in n ways, then you can do task 1 followed by task 2 in $m \times n$ ways.

Degrees of freedom   The number of values in the final calculation of a statistic that are free to vary.

Density curve   Smooth curve that can be used to describe the overall pattern of a distribution. The area under any density curve is always equal to one.

Direction   Used to describe the relationship between two quantitative variables. The direction of the relationship is described as positive or negative.

Discrete random variable   Random variable that can take on only a countable number.

Disjoint (mutually exclusive)   Two events are considered disjoint or mutually exclusive if they cannot occur at the same time.

Distribution   The distribution of a variable tells us what values the variable takes and how often it takes each value.

Double-blind experiment   A double-blind experiment is an experiment in which neither the subjects nor the evaluators know which treatment the subjects have been given.

Empirical Rule (68, 95, 99.7 Rule)   All normal distributions follow the Empirical Rule. That is to say that all normal distributions have: 68% of the observations falling within σ (one standard deviation) of the mean, 95% of the observations falling within 2σ (two standard deviations) of the mean, and 99.7% (almost all) of the observations falling within 3σ (three standard deviations) of the mean.

Experiment   A controlled procedure in which a treatment is imposed on the experimental units or subjects.

Explanatory variable   May be thought of as the independent variable.

Exponential model   Equation used to make predictions when the data are exponential or approximately exponential.

Extrapolation   Extrapolation refers to making predictions about the response variable based on the explanatory variable when the value of the explanatory variable is outside of the domain of the x values. This can be dangerous, as the relationship between x and y can change for extreme values of x.

Event   Subset of a sample space.

Factor(s)   Explanatory variable or variables in an experiment.

Five-number summary   The five-number summary is sometimes used when dealing with skewed distributions. The five-number summary consists of the lowest number, first-quartile (Q_1), median (M), third-quartile (Q_3), and the largest number.

Form   Used to describe the relationship between two quantitative variables. The terms "linear" or "curved" are typically used to describe the form of the relationship.

Geometric distribution   There are four conditions that must be met in order for a distribution to fit a geometric setting. These conditions are:

1. Each observation can be considered a "success" or "failure."
2. The observations must be independent.
3. The probability of success, which we call p, is the same from one trial to the next.
4. The variable that we are interested in is the number of observations it takes to obtain the first success.

Hypothesis test or **test of significance**   Form of statistical inference used when testing a claim that has been made concerning a population.

Independent events   Two events are independent if the occurrence or non-occurrence of one event does not alter the probability of the second event.

Inferential statistics   Conclusions or assumptions about an entire population based on sample data.

Influential observation   Observation that has a dramatic impact on the correlation coefficient and the least squares regression line.

Intersection   The intersection of two events contains all outcomes that belong to both events.

Law of Large Numbers   The Law of Large Numbers states that the long-run relative frequency of repeated, independent trials gets closer to the expected relative frequency once the number of trials increases.

Least Squares Regression Line (LSRL)   The LSRL is fitted to the data by minimizing the sum of the squared residuals. The LSRL equation takes the form of $\hat{y} = a + bx$ where b is the slope and a is the y-intercept. The AP formula sheet uses the form $\hat{y} = b_0 + b_1 x$.

Levels   Actual values for the factors of an experiment. Each factor can have one or more levels.

Linear model   Equation used to make predictions for data that is linear or approximately linear.

Margin of error   When dealing with a one-proportion z-interval, the margin of error is the distance from the endpoints of the confidence interval to the center of the interval, $\hat{p}$. The margin of error is the product of the z* value and the standard error and is affected primarily by the sample size and the z* value (confidence level.) The margin of error for a t-interval is affected in a similar fashion by the sample size and the level of confidence.

Matched-pairs experiment   Type of blocked experiment. The subject can serve as his/her own control, or each subject can be "matched" with another subject on some common characteristic that might affect the experiment. If each subject serves as his/her own control, then the order of the treatment received is randomized. If two subjects are matched, then the treatments are randomly assigned to each of the subjects.

Mean Arithmetic average of the distribution

$$\bar{x} = \frac{x_1 + x_2 + ... + x_n}{n} \quad \text{or} \quad \bar{x} = \frac{\sum x_i}{n}$$

Median Midpoint of the distribution; half of the observations are smaller than the median and half are larger. To find the median: Arrange the data in ascending order (smallest to largest). If there are an odd number of observations, the median is the center data value. If there is an even number of observations, the median is the average of the two middle observations.

Multistage sample Sampling method that combines several different types of sampling. Some national opinion polls are conducted using this method.

Nonresponse Nonresponse can lead to bias when certain individuals who have been selected cannot be reached or choose not to participate in a sample.

Normal distribution Continuous probability distribution. The graph of a normal distribution is bell shaped and follows the 68, 95, 99.7 Rule.

Normal probability plot Used to assess the normality of a population through sample data. A normal probability plot is a scatterplot that graphs a predicted z-score against the value of the variable.

Null hypothesis Hypothesis that states that there is no change or effect in the population. The null hypothesis always includes an equality.

Observational study A type of study where individuals are observed or certain outcomes are measured. No treatment is imposed.

Outcomes Results of the trials of the experiment.

Outliers Values that fall outside the overall pattern of the distribution.

P-value Probability of obtaining a sample statistic as extreme as or more extreme than has been obtained, *given that the null hypothesis is true*. The smaller the p-value, the more evidence to reject the null hypothesis.

Parameter Number that describes some attribute of a population.

Placebo "Dummy pill" used in place of actual medication to help control for the psychological effect of taking medication. The placebo should look and taste just like the treatment being tested.

Placebo effect The placebo effect results when subjects show a response because they believe that they are receiving the actual treatment or medication.

Population All individuals in a particular group of interest.

Power model Equation used to make predictions when data is curved but not exponential.

Power of the test   Probability of rejecting the null hypothesis given that a particular alternative value is true. The power of the test is equal to $1 - \beta$.

Probability   A measure of how likely a particular event is to occur. Probability is always a number between 0 and 1, inclusive.

Quantitative variable   Variable that takes on a numerical value.

Random events   Events that are uncertain but follow a predictable distribution over the long run.

Random variable   Variable from a random experiment that can take on different values. The random variable can be discrete or continuous.

Replication   There are two forms of replication to consider. One type refers to increasing the number of experimental units or subjects, so that it is known that the difference between the experimental group and the control group is really due to the treatment(s) being imposed and not just due to chance. The second type refers to designing an experiment that can be replicated by others doing similar research.

Residual   Difference between the observed value, y, and the predicted value, $\hat{y}$. In other words, *residual = observed – predicted*. Remember that all predicted values are located on the LSRL. A residual can be positive or negative.

Resistant measure   Measurements that resist the influence of extreme data values. Examples of resistant measurements are mean, standard deviation, and the correlation coefficient.

Response bias   Type of bias that results when respondents answer questions in a manner they believe the questioner wants them to answer. Response bias usually occurs due to poorly worded survey questions.

Response variable   May be thought of as the dependent variable.

Robust   The t-procedures are robust. Robust means that the results of our t-interval or t-test would not change very much even though the assumptions of the procedure are violated.

Sample   Part of a population.

Sample space   A list containing all possible outcomes of the experiment.

Sampling distribution   Distribution of the values of the statistic if all possible samples of a given size are taken from the population.

Sampling frame   List of individuals from the entire population from which the sample is drawn.

Sampling variability   Variability that results when repeated samples are taken from the same population. Sampling distributions obtained with smaller sample sizes contain more sampling variability than those obtained from larger samples.

Simple random sample (SRS)   Sample in which every set of n individuals has an equal chance of being chosen.

Simulation   Used to imitate events that involve change behavior.

Single-blind experiment   Experiment in which the subjects do not know which treatment they have received. An experiment is also considered single blind if the subjects do know which treatment they have received but the evaluators do not know which treatment has been given.

Skewed left   The left side of the distribution extends further than the right side, meaning that there are fewer values to the left.

Skewed right   The right side of the distribution extends further than the left side, meaning that there are fewer values to the right.

Spread of the distribution   Refers to the variability of the distribution. Typically, the IQR and/or the variance/standard deviation are used to measure the spread of the distribution.

Standard deviation/variance   Measures the spread of the distribution about the mean. The standard deviation is used to measure spread when the mean is chosen as the measure of center. The standard deviation has the same unit of measurement as the data in the distribution. The variance is the square of the standard deviation and is labeled in units squared.

The formula for variance is:

$$s^2 = \frac{(x_1 - \bar{x})^2 + (x_2 - \bar{x})^2 + ... + (x_n - \bar{x})^2}{n-1} \quad \text{or} \quad s^2 = \frac{\sum (x_i - \bar{x})^2}{n-1}$$

The standard deviation is the square root of the variance.

$$s = \sqrt{\frac{(x_1 - \bar{x})^2 + (x_2 - \bar{x}) + ... + (x_n - \bar{x})}{n-1}} \quad \text{or} \quad s = \sqrt{\frac{\sum (x_i - \bar{x})^2}{n-1}}$$

Standard error   When s is used to estimate σ, the standard deviation of the sampling distribution is called the *standard error* of the sample mean, $\bar{x}$.

Standard normal distribution   Normal distribution with mean of zero and a standard deviation of one. The notation used to denote the standard normal distribution is $N(0,1)$.

Statistic   Number that describes an attribute of a sample.

Statistical inference   Process by which we draw conclusions about an entire population based on sample data.

Stratified Random Sample (SRS)   Sample in which the population is divided into groups that are believed to be similar in some fashion. These homogeneous groups are called *strata*. Within each stratum, an SRS is obtained. These SRS's are then combined to obtain the total sample.

Strength   Used to describe the relationship between two quantitative variables. The strength of the relationship is typically described as weak, moderate, strong, or somewhere in between.

Symmetric distribution   Distribution in which the right and left sides are approximately mirror images of each other.

Systematic sampling   Sampling method in which it is predetermined how the sample will be obtained. For example, you might sample every 25th unit or subject from a given population.

t-distribution   Used when the population standard deviation, σ, is unknown. Like the standard normal distribution, the t-distribution is single-peaked, symmetrical, and bell shaped. As the sample size (n) increases, the variability of the sampling distribution decreases. Thus, as the sample size increases, the t-distributions approach the standard normal model. When the sample size is small, there is more variability in the sampling distribution, and therefore there is more area (probability) under the density curve in the "tails" of the distribution.

Trial   A single attempt of a random event.

Type I error   Occurs when we reject the null hypothesis when, in fact, it is actually correct. The probability of making a type I error is equal to the significance level, ($\alpha - level$) of the test.

Type II error   Occurs when we fail to reject the null hypothesis when, in fact, the null hypothesis is false. The probability of type II error is referred to as β.

Unbiased estimate   An estimate is considered unbiased if it does not systematically tend to overestimate or underestimate the parameter of interest.

Undercoverage   Undercoverage occurs when individuals in the population are excluded in the process of choosing the sample. Undercoverage can lead to bias, so caution must be used.

Union   The union of two events is the event that at least one of the events has occurred.

Voluntary response sample   Sampling method where people respond strictly on a voluntary basis. Voluntary response samples usually include bias, which is referred to as voluntary response bias.

z-score   Standardized normal random variable found by using $z = \dfrac{x - \mu}{\sigma}$.

A z-score is the number of standard deviations a given value is from the mean. Positive z-scores result when an observation is above the mean, and negative z-scores result when an observation is below the mean.

Index